OXFORD SERIES ON NEUTRON SCATTERING IN CONDENSED MATTER

General Editors
S.W. Lovesey, E.W.J. Mitchell

OXFORD SERIES ON NEUTRON SCATTERING IN CONDENSED MATTER

---

1. W.G. Williams: *Polarized neutrons*
2. E. Balcar and S.W. Lovesey: *Theory of magnetic neutron and photon scattering*
3. V.F. Sears: *Neutron optics*
4. M.F. Collins: *Magnetic critical scattering*
5. V.K. Ignatovich: *The physics of ultracold neutrons*
6. Yu. A. Alexandrov: *Fundamental properties of the neutron*
7. P.A. Egelstaff: *An introduction to the liquid state*
8. J.S. Higgins and H.C. Benoît: *Polymers and neutron scattering*
9. H. Glyde: *Excitations in liquid and solid helium*
10. V. Balucani and M. Zoppi: *Dynamic of the liquid state*
11. T.J. Hicks: *Magnetism in disorder*

MAGNETISM IN DISORDER

Trevor J. Hicks

Department of Physics
Monash University, Victoria, Australia

CLARENDON PRESS OXFORD

1995

This book has been printed digitally and produced to a standard design in order to ensure its continuing availability

OXFORD
UNIVERSITY PRESS

Great Clarendon Street, Oxford OX2 6DP

Oxford University Press is a department of the University of Oxford.
It furthers the University's objective of excellence in research, scholarship,
and education by publishing worldwide in

Oxford New York

Auckland Bangkok Buenos Aires Cape Town Chennai
Dar es Salaam Delhi Hong Kong Istanbul Karachi Kolkata
Kuala Lumpur Madrid Melbourne Mexico City Mumbai Nairobi
São Paulo Shanghai Singapore Taipei Tokyo Toronto

with an associated company in Berlin

Oxford is a registered trade mark of Oxford University Press
in the UK and in certain other countries

Published in the United States
by Oxford University Press Inc., New York

Reprinted 2002

A catalogue record for this book is available from the British Library

Library of Congress Cataloging in Publication Data
Hicks, Trevor J.
Magnetism in disorder / Trevor J. Hicks.—1st ed.
(Oxford series on neutron scattering in condensed matter ; 11.)
Includes bibliographical references and index.
1. Nuclear magnetism. 2. Order-disorder models. 3. Neutrons-Scattering.
I. Title. II. Series.
QC761.9.H53 1994 538'.3—dc20 95-6082

ISBN 0-19-851016-0

Dedicated to

E.A. (Bill) Palmer

Executive Secretary of
The Australian Institute of Nuclear Science and Engineering
for thirty years and
committed to the support of science
including
neutron scattering in Australia

Preface

Magnetism in disordered materials is a subject that ranges from the simple isolated defect in a ferromagnet to the description of materials in which there is massive disorder in both atomic arrangement and magnetism such as in the amorphous spin glass materials. It is important that the neutron scattering from these materials be done within a unifying framework. As a result of the magnetic dipole interaction between the electrons in all magnetic materials and neutrons, this is done naturally in terms of the magnetization and the magnetic susceptibility (within the small dipole approximation valid for Fourier components of these quantities with wavelengths longer than atomic dimensions). As most of the region of interest for atomic disorder is interatomic rather than intra-atomic, this seemed a sensible approach. Much of the first chapter is concerned with setting up the magnetic neutron cross sections in these terms.

The gift that the neutron scattering technique hands us is information about the atomic arrangements in disordered materials *at the same time* as the Fourier components of the magnetization or susceptibility are obtained. This, however, highlights another of the emphases of the book: the necessity to distinguish between magnetic and nuclear scattering. I have long been a devotee of the use of polarized neutron techniques and especially that of polarization analysis. In principle, neutron polarization analysis gives us the ability to absolutely distinguish magnetic from nuclear scattering and in many cases this is quite feasible. Consequently, the treatment of scattering with full polarization analysis has been a priority and the description of experiments that incorporate it have been highlighted. The appendix to Chapter 1 presents an analysis of the scattering and analysis of polarized neutrons in the manner of the paper by Moon, Riste, and Koehler (1969)* but is extended to allow for mixed final neutron spin states. I must admit that this gives me some personal satisfaction, as one who was introduced to the subject by Moon, Riste, and Koehler's pioneering paper, and I hope that it provides a new perspective of the subject for those used to this approach.

In most neutron scattering experiments it is the energy and wavevector dependence of the scattering that are observed. An exception is the neutron spin echo experiment where the most natural interpretation of the results is in terms of a real time dependence. Therefore, in the book I do not move from an energy state, Schrödinger representation of the scattering process, which is sufficient for the interpretation of the experiments; rather, time dependences are incorporated in whatever brief description I give of models of the systems under discussion. For instance, the Green function models of the excitations in atomically disordered antiferromagnets in Chapter 5, are related to the imaginary

* Moon, R.M., Riste, T. and Koehler, W.C. (1969). *Phys. Rev.* **181**, 920.

part of the wavevector and energy dependent susceptibility, which have already been related to the neutron cross section.

I have also avoided giving too detailed a description of models that apply to particular systems. If I have described a model it has generally been phenomenological. In this way I have hoped to make the book accessible to a wider audience while at the same curtailing its length. The disadvantage of this is that while a qualitative description is usually given the reader will have to probe further for a detailed microscopic description of individual systems. For better or worse I regarded my task as dealing primarily with the particular aspects of neutron scattering from magnetic systems with atomic disorder. However, I am strongly of the opinion that it is important to make the connection between neutron scattering and other probes of these materials. I have therefore pointed out the quantitative comparison of neutron scattering cross sections with bulk magnetic parameters at every opportunity, a task made easy because of the already established connections with the magnetization and the susceptibility.

I am grateful to Professor S.W. Lovesey for asking me to write this book. I am afraid that my initial reaction was not positive but after reflection I realized that I would like to do it. Now I am pleased that I finally agreed. The experience has been a good one and has given me the opportunity to integrate many of the thoughts that I had previously used in papers and lectures. Ideas become experiences within a group context and I am grateful to my students and associates in this department, some of whose work is described in this book. Among those were J.H. Smith, N. Ahmed, S.J. Campbell, J.R. Davis, R. Cywinski, P. Wells, O. Moze, T.M. Harders, P. Gibbs, and E. Bakshi. I am also very grateful to Mr S. McCausland who redrew the figures for this book.

I wrote the bulk of the book during 1991 while a guest of the Hahn–Meitner Institut in Berlin and the Oak Ridge National Laboratory, Tennessee. For this I am very thankful to Professor F. Mezei, and to Dr J.W. Cable and Dr R.M. Moon. In both laboratories I was free to write and to interact with colleagues, and do experiments. I would particularly like to thank Dr J.W. Cable for having a close look at many of the chapters.

Even a modest book like this requires a lot of time and effort—time mostly robbed from one's family, and effort mostly robbed from homely chores. To my wife, Dorothy, and my daughter, Julia, I must express my great gratitude for their patience and love in this time. It took longer than I expected!

Glen Iris, Victoria T.J.H.
December 1993

Acknowledgements

I am grateful to the following publishers and institutions for permission to use their illustrations as figures in this book: The American Institute of Physics, The American Physical Society, Les Éditions de Physique, Elsevier Science Publishers, The Institut Laue–Langevin, IOP Publishing, Il Nuovo Cimento, Pergamon Press, Polska Akademia Nauk, Springer-Verlag, Taylor and Francis Ltd, and USERDA with Oak Ridge National Laboratory.

Contents

Glossary of symbols xiii

1 Interactions 1

1.1 The nuclear and magnetic potentials 1
1.2 Neutron spatial states 2
1.3 Neutron spin states 3
1.4 The elastic cross section 4
1.5 Inelastic scattering and the generalized susceptibility 6
1.6 Magnetic quasielastic scattering 9
Appendix: Neutron spin transitions 10
References 14

2 Dilute ferromagnetic alloys 15

2.1 Defect cross sections for ferromagnetic alloys 15
2.2 The GLOPPER 17
2.3 Dilute ferromagnetic alloys 18
2.4 Data from dilute nickel alloys 19
2.5 Dilute cobalt alloys 21
2.6 Dilute iron alloys 22
2.7 Dilute impurities in exchange enhanced hosts 24
References 25

3 Concentrated ferromagnetic alloys 27

3.1 Atomic short range order 27
3.2 Macroscopic limits 29
3.3 Models for concentrated ferromagnetic alloys 31
3.4 The magnetic environment model 36
3.5 Application of the magnetic environment model 39
3.6 The average moment distribution in ferromagnetic alloys 46
References 49

4 Defect scattering from antiferromagnetic alloys 50

4.1 Polarization analysis instruments for diffuse scattering 51
4.2 Models for antiferromagnets 54
4.3 γ-Mn–Ni alloys 58
4.4 Antiferromagnetic Mn–Cu alloys 65

4.5 Nonmetallic antiferromagnetic alloys 66
References 70

5 Excitations **71**
5.1 Average magnons in ferromagnetic alloys 71
5.2 The cluster model for antiferromagnetic insulators 74
5.3 Modes at general wavevectors 76
5.4 Critical scattering 80
References 84

6 Spin glasses **86**
6.1 Spin correlations and susceptibility 86
6.2 The small κ susceptibility of dilute Cu–Mn spin glasses 89
6.3 Larger κ features in Cu–Mn alloys 92
6.4 Au–Fe spin glasses 94
6.5 Dynamics of Cu–Mn spin glasses 97
6.6 Other spin glasses 99
6.7 Re-entrant spin glasses 103
6.8 Spin echo spectroscopy 109
6.9 Spin glass dynamics 110
6.10 Magnons in re-entrant spin glasses 114
References 119

7 Amorphous magnets **121**
7.1 The structure of ferromagnetic glasses 121
7.2 Magnons in amorphous ferromagnets 125
7.3 Amorphous spin glasses 131
7.4 Amorphous re-entrant spin glasses 134
References 140

Name index **143**

Subject index **145**

Glossary of symbols

α	a Cartesian direction; fraction of neutron spin up amplitude
$\alpha(\boldsymbol{R})$	Cowley short range order parameter
β	the inverse of Boltzmann's constant times temperature; a Cartesian direction; fraction of neutron spin down amplitude
γ	neutron gyromagnetic ratio; a Cartesian direction; spin–spin correlation parameter
$\zeta(\boldsymbol{R})$	initial magnetic susceptibility at site $\boldsymbol{R}$
$\boldsymbol{\kappa}$	scattering vector
$\tilde{\boldsymbol{\kappa}}$	unit scattering vector
$\boldsymbol{\mu}$	neutron magnetic moment
$\boldsymbol{\mu}(\boldsymbol{R})$	atomic magnetic moment at site $\boldsymbol{R}$
μ_B	Bohr magneton
μ_N	nuclear magneton
ν	staggered atomic moment in an antiferromagnet
$\nu(\boldsymbol{R})$	sum of nearby site occupation operators
$\boldsymbol{\rho}$	vector difference between two lattice sites
σ	cross section
$\boldsymbol{\sigma}$	neutron spin angular momentum
$\Sigma(\omega)$	effective site potential in CPA
$\Sigma(z)$	self energy as a function of the complex integration variable
$\boldsymbol{\tau}$	antiferromagnetic reciprocal lattice vector
$\Phi(\boldsymbol{\kappa})$	Fourier transform of $\phi(\boldsymbol{R})$
$\phi(\boldsymbol{R})$	magnetic moment disturbance at $\boldsymbol{R}$
$X(t)$	time dependent magnetic response function
χ''	imaginary part of magnetic susceptibility
χ'	real part of magnetic susceptibility
ω	angular frequency
Ω	solid angle
$A(\boldsymbol{\kappa})$	sum of nuclear spin and magnetic moment amplitudes for neutron scattering
b	nuclear scattering length
B	nuclear neutron spin interaction parameter; total specimen scattering length; moment saturation parameter
$\mathbf{B}(\boldsymbol{r})$	magnetic induction
B_1	parameter for moment saturation at a site occupied by an atom of type 1

c	velocity of light
C	fractional atomic concentration
D	magnon stiffness
$D(\boldsymbol{\kappa})$	Fourier transform of spatial fluctuations in atomic occupation
e	electron charge
E	energy
f	atomic magnetic form factor
$F(\boldsymbol{\kappa})$	Fourier transform of spatial fluctuations in atomic moment
F_{AB}	partial structure factor for atoms of type A and B
g	Landé factor for an atomic moment
$G(\omega)$	Green function
H	magnetic field intensity
$h(\boldsymbol{R})$	exchange field at $\boldsymbol{R}$
$\boldsymbol{I}$	nuclear angular momentum
I	neutron intensity
$I(\boldsymbol{\kappa})$	Fourier transform of $\mathscr{I}$
$\mathscr{I}(\boldsymbol{R}_i - \boldsymbol{R}_j)$	function connecting a moment at j with an exchange field at i
$J(\boldsymbol{\kappa})$	Fourier transform of exchange interaction
$j(\boldsymbol{r})$	exchange interaction
$\boldsymbol{k}$	incident neutron wavevector
k_B	Boltzmann's constant
$\boldsymbol{k}'$	scattered neutron wavevector
m	neutron mass
$m(\boldsymbol{\kappa})$	ratio of magnetic to atomic spatial fluctuations
$\mathbf{M}(\boldsymbol{\kappa})$	Fourier component of the magnetization
$\boldsymbol{M}(\boldsymbol{\kappa})$	Fourier component of the magnetic moment
m_e	electron mass
n	precession number
N	number of sites in sample
$\boldsymbol{P}$	incident neutron polarization
$P(\boldsymbol{\kappa})$	Fourier transform of the site occupancy operator p_i
p_i	occupation operator for site i
p_q	probability of target being in state q
$\boldsymbol{P}'$	scattered neutron polarization
q	initial target state
$\boldsymbol{q}$	magnon wavevector
q'	final target state
$\boldsymbol{r}$	real space position
$\boldsymbol{R}$	atom position
s	initial neutron spin state
S	neutron or atomic spin operator or expectation value
$S(\boldsymbol{\kappa})$	Fourier transform of the Cowley atomic short range order parameters
s'	final neutron spin state

t	time; single atom T-matrix
T	temperature
T	transition probability between initial and final neutron spin states
U	neutron transition amplitude from z polarization to other polarizations
v	velocity
V	volume
$\mathrm{V}(\boldsymbol{r})$	potential energy
V_B	Brillouin zone volume
$W_{qq'}$	transition probability between target states q and q'
z	atomic coordination
$\hbar$	Planck's constant divided by 2π

1

Interactions

The magnetic properties of disordered solids are rich and varied and no one technique is sufficient to probe the whole range of behaviour. Nevertheless, neutron scattering comes closest to universal applicability. Spatial inhomogeneity is an important aspect of disordered materials. For some techniques spatial information can only be inferred indirectly from the data, for instance by assuming that the largest effects are most probably from first neighbour atoms. Spatial information is built directly into neutron scattering experiments because of the neutron wavelength. The accessible length scale is from atomic dimensions to distances of hundreds of Ångstroms; a very useful range for disorder at the atomic level.

The direct sensing of the magnetic induction is the other great advantage of neutron scattering for magnetic materials. There are other techniques, some of which may, in particular circumstances, be more sensitive, but for which there is a long chain of modelling of the interaction before a magnetic property, like a magnetic moment value, can be inferred. One of the purposes of this chapter will be to draw out the direct relationship between the neutron cross section and the magnetization and susceptibility of magnetic materials.

One way in which the magnetic scattering from disordered materials differs from that from well-ordered materials is in the sharpness of the features in reciprocal space and energy. There are many important values, like the average moment, which can be directly extracted from the Bragg scattering, but the details of the effects of disorder are contained in the diffuse scattering. For this reason it can be important to be able to unequivocally separate out what is magnetic scattering from the rest. Having done so, much of what is left is due to nuclear disorder scattering which characterizes the atomic disorder. There are various means of making this separation, most of which involve the use of polarized neutrons, and so much of this chapter explores the sometimes complex relationships governing the scattering of polarized neutrons.

1.1 THE NUCLEAR AND MAGNETIC POTENTIALS

The interaction of the neutron with the nucleus can be parameterized by the potential

$$\hat{\mathrm{V}}(\boldsymbol{r}) = \left(\frac{2\pi\hbar^2}{m}\right)\left(b + B\hat{\boldsymbol{I}}\cdot\hat{\boldsymbol{\sigma}}\right)\delta(\boldsymbol{r}-\boldsymbol{R}) \tag{1.1}$$

where the nucleus is located at $\boldsymbol{R}$. The interaction of the neutron with the magnetization of magnetic materials is via the magnetic dipole moment of the neutron. The magnetic moment operator of the neutron can be written

$$\hat{\boldsymbol{\mu}} = \gamma\mu_N\hat{\boldsymbol{\sigma}} \tag{1.2}$$

where $\hat{\boldsymbol{\sigma}}$ is the Pauli spin matrix for spin 1/2 particles, μ_N is the nuclear magneton, and $\gamma = -1.91$ is the gyromagnetic ratio for the neutron. The interaction of a neutron with a magnetic induction $\mathbf{B}(\boldsymbol{r})$ is given by

$$-\hat{\boldsymbol{\mu}}\cdot\mathbf{B}(\boldsymbol{r}) = -\gamma\mu_N\hat{\boldsymbol{\sigma}}\cdot\mathbf{B}(\boldsymbol{r}). \tag{1.3}$$

This approximation in which the sum of all the magnetic dipoles in the material is represented by the magnetic induction $\mathbf{B}(\boldsymbol{r})$ is useful when the variations being probed have a scale much larger than the individual dipoles. In this book we shall be generally concerned with variations of $\mathbf{B}(\boldsymbol{r})$ which are extra-atomic rather than intra-atomic and therefore of a larger scale than atomic orbital magnetic dipole moments.

The total interaction of the neutron at a point $\boldsymbol{r}$ in a target is

$$\hat{\mathrm{V}}(\boldsymbol{r}) = \left(\frac{2\pi\hbar^2}{m}\right)\left[b\delta(\boldsymbol{r}-\boldsymbol{R}) + \left\{\boldsymbol{B}\hat{\boldsymbol{I}}\delta(\boldsymbol{r}-\boldsymbol{R}) - \frac{1}{4\pi}\left(\frac{e\gamma}{\hbar c}\right)\hat{\mathbf{B}}(\boldsymbol{r})\right\}\cdot\hat{\boldsymbol{\sigma}}\right] \tag{1.4}$$

where the terms have been grouped according to the interaction with the neutron spin, and smaller terms such as those involving the velocity of the neutron and the electric field gradient (Lovesey 1984*b*) have been ignored.

1.2 NEUTRON SPATIAL STATES

In the first Born approximation the neutron cross section is given by

$$\left(\frac{\mathrm{d}\sigma}{\mathrm{d}\Omega\mathrm{d}E}\right)_{k,k'} = \frac{k'}{k}\left(\frac{m}{2\pi\hbar^2}\right)^2\sum_{q'}\sum_{q} p_q\left|\langle \boldsymbol{k}'s'q'|\hat{V}(\boldsymbol{r})|\boldsymbol{k}sq\rangle\right|^2\delta(\hbar\omega - E_{q'} + E_q), \tag{1.5}$$

(Lovesey 1984*a*) where the integral is over the interaction volume V, and the neutron goes from spatial plane wave state $\boldsymbol{k}$ to plane wave state $\boldsymbol{k}'$, and from spin state s to s'. At the same time the specimen goes from state q to state q'.

We firstly consider the transition between neutron spatial states. The matrix element may be written

$$\langle s'q'|\int \mathrm{d}\boldsymbol{r}\left(\frac{2\pi\hbar^2}{m}\right)\left\{\hat{b}\delta(\boldsymbol{r}-\boldsymbol{R}) + \boldsymbol{B}\hat{\boldsymbol{\sigma}}\cdot\hat{\boldsymbol{I}}\delta(\boldsymbol{r}-\boldsymbol{R}) - \frac{1}{4\pi}\left(\frac{e\gamma}{\hbar c}\right)\hat{\boldsymbol{\sigma}}\cdot\hat{\boldsymbol{B}}(\boldsymbol{r})\right\}\mathrm{e}^{\mathrm{i}(\boldsymbol{k}-\boldsymbol{k}')\cdot\boldsymbol{r}}|sq\rangle \tag{1.6}$$

after inserting the plane wave states for $\boldsymbol{k}$ and $\boldsymbol{k}'$, and this becomes

$$\left(\frac{2\pi\hbar^2}{m}\right)\left\{\sum_{R}\left[\hat{b}(R)+B(R)\hat{I}(R)\cdot\hat{\sigma}\right]\mathrm{e}^{\mathrm{i}\kappa\cdot R}-\frac{V}{4\pi}\left(\frac{e\gamma}{\hbar c}\right)\hat{\sigma}\cdot\hat{\mathbf{B}}(\kappa)\right\}, \tag{1.7}$$

where $\kappa = k' - k$, and

$$\hat{\mathbf{B}}(\kappa)=\frac{1}{V}\int \mathrm{d}r\;\hat{\mathbf{B}}(r)\mathrm{e}^{-\mathrm{i}\kappa\cdot r}, \tag{1.8}$$

with V the specimen volume. In the absence of an external field $\mathbf{B}(\kappa)$ is

$$\hat{\mathbf{B}}(\kappa)=4\pi\hat{\mathrm{M}}_{\perp}(\kappa), \tag{1.9}$$

where $\mathbf{M}(\kappa)$ is the Fourier transform of the magnetization distribution. The components of $\mathbf{M}(\kappa)$ parallel to κ are not reflected in $\mathbf{B}(\kappa)$ because of the requirement that $\nabla\cdot\mathbf{B} = 0$.

The matrix element can then be written

$$\left(\frac{2\pi\hbar^2}{m}\right)\left\langle s'q'\right|b(\kappa)+\hat{A}(\kappa)\cdot\hat{\sigma}\left|qs\right\rangle, \tag{1.10}$$

in which $b(\kappa)$ is the Fourier transform of the scattering length density, and

$$\hat{A}(\kappa)=\sum_{R}B(R)\hat{I}(R)\,\mathrm{e}^{\mathrm{i}\kappa\cdot R}+\left(\frac{e\gamma}{\hbar c}\right)\hat{M}_{\perp}(\kappa), \tag{1.11}$$

with $\hat{M}(\kappa)$ the Fourier transform of the total magnetic moment operator for the specimen.

1.3 NEUTRON SPIN STATES

At this point we need to evaluate the matix elements between the spin states s and s'. This can be done using the conventional density matrix formalism (Blume 1963; Lovesey 1984*b*) or by the method outlined in the Appendix (at the end of this chapter) where the results are obtained using the neutron spin states explicitly. The expressions obtained are firstly for the transition probability T from a polarized neutron spin state to all final spin states

$$\mathrm{T}=\left(\frac{2\pi\hbar^2}{m}\right)^2\left[b^*(\kappa)b(\kappa)+A^*(\kappa)\cdot A(\kappa)+b^*(\kappa)A(\kappa)\cdot P+\right.$$

$$\left.A^*(\kappa)\cdot P\,b(\kappa)+\mathrm{i}\left\{A^*(\kappa)\times A(\kappa)\right\}\cdot P\right], \tag{1.12}$$

where for instance $b^*(\kappa)$ has been written for $<q|\hat{b}^*(\kappa)|q'>$, and a similar nomeclature has been used for the other transition amplitudes. This corresponds to an experiment in which there is an incident neutron polarization but no analysis for final polarization. The second expression is for the polarization of the scattered neutrons $\boldsymbol{P}'$ given an incident beam polarization $\boldsymbol{P}$

$$\begin{aligned}\boldsymbol{P}'\mathrm{T} = \left(\frac{2\pi\hbar^2}{m}\right)^2 &\left[b^*(\boldsymbol{\kappa})b(\boldsymbol{\kappa})\boldsymbol{P} + b^*(\boldsymbol{\kappa})\boldsymbol{A}(\boldsymbol{\kappa}) + \boldsymbol{A}^*(\boldsymbol{\kappa})b(\boldsymbol{\kappa}) + \right.\\ &\mathrm{i}b^*(\boldsymbol{\kappa})\{\boldsymbol{A}(\boldsymbol{\kappa})\times\boldsymbol{P}\} + \mathrm{i}\{\boldsymbol{P}\times\boldsymbol{A}^*(\boldsymbol{\kappa})\}b(\boldsymbol{\kappa}) - \{\boldsymbol{A}^*(\boldsymbol{\kappa})\cdot\boldsymbol{A}(\boldsymbol{\kappa})\}\boldsymbol{P} + \\ &\left.\{\boldsymbol{A}^*(\boldsymbol{\kappa})\cdot\boldsymbol{P}\}\boldsymbol{A}(\boldsymbol{\kappa}) + \boldsymbol{A}^*(\boldsymbol{\kappa})\{\boldsymbol{A}(\boldsymbol{\kappa})\cdot\boldsymbol{P}\} - \mathrm{i}\{\boldsymbol{A}^*(\boldsymbol{\kappa})\times\boldsymbol{A}(\boldsymbol{\kappa})\}\right],\end{aligned} \tag{1.13}$$

where it can be seen that different combinations of nuclear and magnetic scattering amplitudes affect the final polarization in different ways, thus providing a means for their separation. In addition, there are terms which require no polarization of the incident beam for a final polarization. These are the cross terms between nuclear scattering amplitude, and vector magnetic and nuclear amplitudes, and that arising from a purely vector amplitude associated with a definite direction such as a magnetic spiral structure. These are the terms used to produce polarized beams by reflection or scattering from magnetic materials.

1.4 THE ELASTIC CROSS SECTION

We shall firstly consider the elastic scattering from a system in which the scattering from nuclear spins is zero. In elastic scattering the target is left in the same state so that $q' = q$, and all the transition amplitudes in eqns (1.12), (A.13)and (1.13), (A.14) become expectation values. The unanalysed cross section is

$$\begin{aligned}\frac{\mathrm{d}\sigma}{\mathrm{d}\Omega} = &\overline{\langle\hat{b}^*(\boldsymbol{\kappa})\rangle\langle\hat{b}(\boldsymbol{\kappa})\rangle} + \left(\frac{e\gamma}{\hbar c}\right)^2 \overline{\langle\hat{\boldsymbol{M}}_\perp^*(\boldsymbol{\kappa})\rangle\cdot\langle\hat{\boldsymbol{M}}_\perp(\boldsymbol{\kappa})\rangle}\\ &-\frac{e\gamma}{\hbar c}\left\{\overline{\langle\hat{b}^*(\boldsymbol{\kappa})\rangle\langle\hat{\boldsymbol{M}}_\perp(\boldsymbol{\kappa})\rangle}\cdot\boldsymbol{P} + \overline{\langle\hat{\boldsymbol{M}}_\perp^*(\boldsymbol{\kappa})\rangle\cdot\boldsymbol{P}\langle\hat{b}(\boldsymbol{\kappa})\rangle}\right\}\\ &+\left(\frac{e\gamma}{\hbar c}\right)^2 \mathrm{i}\overline{\left\{\langle\hat{\boldsymbol{M}}_\perp^*(\boldsymbol{\kappa})\rangle\times\langle\hat{\boldsymbol{M}}_\perp(\boldsymbol{\kappa})\rangle\right\}}\cdot\boldsymbol{P},\end{aligned} \tag{1.14}$$

where the bars are the thermal averages over the states q. Similarly, the final polarization is given by

$$\begin{aligned}\boldsymbol{P}'\frac{\mathrm{d}\sigma}{\mathrm{d}\Omega} = &\overline{\langle b^*(\boldsymbol{\kappa})\rangle\langle b(\boldsymbol{\kappa})\rangle}\boldsymbol{P} - \left(\frac{e\gamma}{\hbar c}\right)\left[\overline{\langle b^*(\boldsymbol{\kappa})\rangle\langle\hat{\boldsymbol{M}}_\perp(\boldsymbol{\kappa})\rangle} + \right.\\ &\overline{\langle\hat{\boldsymbol{M}}_\perp^*(\boldsymbol{\kappa})\rangle\langle b(\boldsymbol{\kappa})\rangle} + \mathrm{i}\overline{\langle b^*(\boldsymbol{\kappa})\rangle(\langle\hat{\boldsymbol{M}}_\perp(\boldsymbol{\kappa})\rangle\times\boldsymbol{P})} + \\ &\left.\mathrm{i}\overline{(\boldsymbol{P}\times\langle\hat{\boldsymbol{M}}_\perp^*(\boldsymbol{\kappa})\rangle)\langle b(\boldsymbol{\kappa})\rangle}\right\} - \left(\frac{e\gamma}{\hbar c}\right)^2\overline{\left\{(\langle\hat{\boldsymbol{M}}_\perp^*(\boldsymbol{\kappa})\rangle\cdot\langle\hat{\boldsymbol{M}}_\perp(\boldsymbol{\kappa})\rangle)\boldsymbol{P} - \right.}\end{aligned}$$

$$\overline{\left(\left\langle \hat{\boldsymbol{M}}_\perp^*(\boldsymbol{\kappa})\right\rangle \cdot \boldsymbol{P}\right)\left\langle \hat{\boldsymbol{M}}_\perp\right\rangle} - \overline{\left\langle \hat{\boldsymbol{M}}_\perp^*(\boldsymbol{\kappa})\right\rangle\left(\left\langle \hat{\boldsymbol{M}}_\perp(\boldsymbol{\kappa})\right\rangle \cdot \boldsymbol{P}\right)} + \tag{1.15}$$

$$\mathrm{i}\overline{\left(\left\langle \hat{\boldsymbol{M}}_\perp^*(\boldsymbol{\kappa})\right\rangle \times \left\langle \hat{\boldsymbol{M}}_\perp(\boldsymbol{\kappa})\right\rangle\right)}\Bigg\}.$$

This latter expression appears complicated but in practice one is analysing along Cartesian directions to isolate particular parts of the cross section, and the very complexity of the relationship is an aid to the separation.

For ferromagnetic materials there are two ways in which the above equations have been used to isolate magnetic cross sections or amplitudes. The first uses an unpolarized beam and manipulates the direction of the magnetization with respect to the scattering vector by means of an applied field. Because the initial polarization is zero and the material is a collinear ferromagnet only the first two terms of eqn (1.14) are nonzero. The second is also zero if the magnetization is entirely along the scattering vector and assumes full value if aligned perpendicularly to the scattering vector. In this way the magnetic cross section can be switched on and off by the appropriate application of a magnetic field. In zero field the various ferromagnetic domains relax back to their individual magnetization directions and provide a nonzero magnetic cross section which depends on their zero field configuration. For cubic materials this provides two-thirds of the maximum magnetic cross section. The second common way isolates the magnetic amplitude of a collinear ferromagnet if the nuclear amplitude is known. By using a polarized beam all but the third and fourth terms of eqn (1.14) can eliminated by subtracting the intensities obtained with initial polarization along and opposite to the magnetization direction. The result is then a product of the magnetic and nuclear amplitudes. This method can be very sensitive if the nuclear scattering amplitude is large.

For nonferromagnetic materials, except single domain spiral structures, no isolation of the magnetic cross section is possible without analysis of the final polarization. This is because these materials have no unique direction. Collinear antiferromagnets merely have an axis. The principal method for the isolation of the magnetic cross section for these nonferromagnetic materials rests on the purely magnetic terms of eqn (1.15) which depend on the initial polarization. Investigation of these three terms will show that if the initial polarization is directed along the scattering vector the polarization is reversed on scattering. There is only one other source of polarization reversal for this geometry; that due to the nuclear spin scattering (1.20).

In most neutron experiments the nuclear spin scattering appears as elastic scattering, although strictly it it is entirely inelastic in the absence of nuclear order and in zero field. This is because the energy states involving the nuclear spins are of very low energy compared with the excitations of the lattice or of the electronic magnetic system. The nuclear system is almost always at temperatures and in magnetic fields in which there is no correlation between spins on different sites and in which the thermally averaged spin at any site is zero. The different types of terms containing the nuclear spin operator which enter the expressions for the transition probabilities (1.12), (A.13), and (1.13), (A.14) are $b^*(\boldsymbol{\kappa})I(\boldsymbol{\kappa})$, $M^*_{\perp x}(\boldsymbol{\kappa})I_y(\boldsymbol{\kappa})$, $I^*_x(\boldsymbol{\kappa})I_y(\boldsymbol{\kappa})$, and $I^*_x(\boldsymbol{\kappa})I_x(\boldsymbol{\kappa})$. The first of these terms for strictly elastic scattering is

$$\overline{\langle \hat{b}^*(\boldsymbol{\kappa})\rangle}\;\overline{\langle \hat{I}(\boldsymbol{\kappa})\rangle} \tag{1.16}$$

in which the thermal averages of the two expectation values have been taken separately because there is no correlation between them. In fact the thermal average of $\langle I(\boldsymbol{\kappa})\rangle = 0$ so that this term is zero. The result is the same if we take any common final state of the two operators. The second term can be treated in the same way and is also zero. For the terms of the third type we can include all the final states of the nuclear system in our close to elastic scattering regime and sum over final states $q'$by closure so that

$$\sum_{q,q'} p_q \langle q|\hat{\boldsymbol{I}}^*_x(\boldsymbol{\kappa})|q'\rangle\langle q'|\hat{\boldsymbol{I}}_y(\boldsymbol{\kappa})|q\rangle = \overline{\langle \hat{I}^*_x(\boldsymbol{\kappa})\hat{I}_y(\boldsymbol{\kappa})\rangle} \tag{1.17}$$

$\boldsymbol{I}_x$ and $\boldsymbol{I}_y$ are not correlated except on the same site where they do not commute, so that terms of this type are also zero. The fourth type of term can be treated in a similar manner, but in this case

$$\overline{\langle \hat{\boldsymbol{I}}^*_x(\boldsymbol{\kappa})\hat{\boldsymbol{I}}_x(\boldsymbol{\kappa})\rangle} = \frac{N}{3}\overline{\boldsymbol{I}(\boldsymbol{R})(\boldsymbol{I}(\boldsymbol{R})+1)} \tag{1.18}$$

as the same components of the nuclear spin operator do commute on the same site. The first bar is a thermal average and the bar over the right-hand side is an average over sites. This treatment adds a term

$$\frac{d\sigma}{d\Omega} = N\overline{B^2(\boldsymbol{R})\boldsymbol{I}(\boldsymbol{R})[\boldsymbol{I}(\boldsymbol{R})+1]} \tag{1.19}$$

to the unanalysed cross section in eqn (1.14), and a term

$$\boldsymbol{P}'\frac{d\sigma}{d\Omega} = -\frac{N}{3}\overline{B^2(\boldsymbol{R})\boldsymbol{I}(\boldsymbol{R})[\boldsymbol{I}(\boldsymbol{R})+1]}\,\boldsymbol{P} \tag{1.20}$$

to the final polarization in eqn (1.15).

1.5 INELASTIC SCATTERING AND THE GENERALIZED SUSCEPTIBILITY

The inelastic scattering is directly related to the magnetic susceptibility of a material in a similar manner to the relationship of the elastic scattering with the thermal average of the expectation value of the magnetic moment. This is important, especially for disordered materials because in this case it may be more difficult to calculate the scattering from microscopic considerations. In many cases the scattering results may be compared with susceptibility results in a model free manner and microscopic models may be considered in this wider context. We therefore follow the connection between the neutron scattering cross section and the generalized susceptibility.

The uniform frequency dependent susceptibility, $\chi_{\alpha\alpha}(\omega)$, is defined as

$$\boldsymbol{M}_{\alpha}(t) = \tfrac{1}{2} V\left(\chi_{\alpha\alpha}(\omega) H_{0_\alpha} \mathrm{e}^{-\mathrm{i}\omega t} + \chi^{*}_{\alpha\alpha}(\omega) H^{*}_{0_\alpha} \mathrm{e}^{\mathrm{i}\omega t}\right), \tag{1.21}$$

where $M_{\alpha}(t)$ is the total moment resulting from the impressed oscillating field

$$H_{\alpha}(t) = \tfrac{1}{2}\left(H_{0_\alpha} \mathrm{e}^{-\mathrm{i}\omega t} + H^{*}_{0_\alpha} \mathrm{e}^{\mathrm{i}\omega t}\right), \tag{1.22}$$

and V is the specimen volume. The lag in response is described by the relative magnitudes of the real and imaginary parts of

$$\chi_{\alpha\alpha}(\omega) = \chi'_{\alpha\alpha}(\omega) + \mathrm{i}\chi''_{\alpha\alpha}(\omega), \tag{1.23}$$

and the rate of energy gain in the specimen is

$$\frac{\mathrm{d}E}{\mathrm{d}t} = -\boldsymbol{M}_{\alpha} \frac{\mathrm{d}H_{\alpha}}{\mathrm{d}t}, \tag{1.24}$$

so that the average rate of energy gain is

$$\overline{\frac{\mathrm{d}E}{\mathrm{d}t}} = \tfrac{1}{2} V\, H^{*}_{0_\alpha} H_{0_\alpha}\, \omega \chi''_{\alpha\alpha}(\omega). \tag{1.25}$$

The power can also be calculated within first order perturbation theory as follows

$$\overline{\frac{\mathrm{d}E}{\mathrm{d}t}} = \sum_{q} \sum_{q'} p_q W_{qq'}\left(E_{q'} - E_q\right), \tag{1.26}$$

with p_q the probability that the system is in state q, and with $E_{q'}$ and E_q the energies of the two states. The transition probability is given by

$$W_{qq'} = \frac{\pi}{\hbar} H^{*}_{0\alpha} H_{0_\alpha} \left|\langle q' | M_{\alpha} | q\rangle\right|^{2} \delta\left(\hbar\omega - E_{q'} + E_q\right). \tag{1.27}$$

But

$$\left|\langle q | \boldsymbol{M}_{\alpha} | q'\rangle\right|^{2} = \left|\langle q' | \boldsymbol{M}_{\alpha} | q\rangle\right|^{2}, \tag{1.28}$$

and the net rate in energy gain, i.e. absorption minus stimulated emission, is thus

$$\begin{aligned}\overline{\frac{\mathrm{d}E}{\mathrm{d}t}} &= \tfrac{1}{2} \sum_{q} \sum_{q \neq q'} \frac{\pi}{\hbar} \boldsymbol{H}_{0_\alpha}{}^{*} \boldsymbol{H}_{0_\alpha} \left|\langle q' | \boldsymbol{M}_{\alpha} | q\rangle\right|^{2} p_q \left\{1 - \frac{\exp(-\beta E')}{\exp(-\beta E)}\right\} \left(E_{q'} - E_q\right) \delta\left(\hbar\omega - E_{q'} + E_q\right) \\ &= \tfrac{1}{2} \sum_{q} \sum_{q' \neq q} \pi \boldsymbol{H}_{0_\alpha}{}^{*} \boldsymbol{H}_{0_\alpha} \left|\langle q' | \boldsymbol{M}_{\alpha} | q\rangle\right|^{2} p_q \omega \left(1 - \mathrm{e}^{-\beta\hbar\omega}\right) \delta\left(\hbar\omega - E_{q'} + E_q\right),\end{aligned} \tag{1.29}$$

with 1/2 compensating for double counting in the sum over the states, and omitting $q'=q$ because it involves no change in energy. Finally, by equating energy rates calculated in the two ways we have

$$\frac{\chi''_{\alpha\alpha}(\omega)}{\pi\left(1-e^{-\beta\hbar\omega}\right)}=\frac{1}{V}\sum_{q}\sum_{q'\neq q}p_q\left|\langle q'|M_\alpha|q\rangle\right|^2\delta\left(\hbar\omega-E_{q'}+E_q\right). \tag{1.30}$$

The cross sections may be obtained by inserting the transition probabilities from (1.12), (A.13) and (1.13), (A.14) into the expression for the cross section (1.5). The cross section without analysis and only including the purely magnetic terms is

$$\left(\frac{d^2\sigma}{d\Omega dE}\right)_{k,k'}=\frac{k'}{k}\left(\frac{m}{2\pi\hbar^2}\right)^2\sum_{q}\sum_{q'\neq q}p_q\,\mathrm{T}\;\delta\left(\hbar\omega-E_{q'+}E_q\right), \tag{1.31}$$

with

$$\mathrm{T}=\left(\frac{2\pi\hbar^2}{m}\right)^2\left(\frac{e\gamma}{\hbar c}\right)^2\left[\boldsymbol{M}_\perp^*(\boldsymbol{\kappa})\cdot\boldsymbol{M}_\perp(\boldsymbol{\kappa})+\mathrm{i}\left\{\boldsymbol{M}_\perp^*(\boldsymbol{\kappa})\times\boldsymbol{M}_\perp(\boldsymbol{\kappa})\right\}\cdot\boldsymbol{P}\right], \tag{1.32}$$

where as before the amplitude $M(\boldsymbol{\kappa}) = \langle q'|M(\boldsymbol{\kappa})|q\rangle$. The final polarization can be obtained from

$$\boldsymbol{P}'\left(\frac{d^2\sigma}{d\Omega dE}\right)_{k,k'}=\frac{k'}{k}\left(\frac{m}{2\pi\hbar^2}\right)^2\sum_{q}\sum_{q'\neq q}p_q\boldsymbol{P}'\mathrm{T}\delta\left(\hbar\omega-E_{q'}+E_q\right), \tag{1.33}$$

$$\boldsymbol{P}'\mathrm{T}=\ -\left(\frac{2\pi\hbar^2}{m}\right)^2\left(\frac{e\gamma}{\hbar c}\right)^2\left[\left\{\boldsymbol{M}_\perp^*(\boldsymbol{\kappa})\cdot\boldsymbol{M}_\perp(\boldsymbol{\kappa})\right\}\boldsymbol{P}-\left\{\boldsymbol{M}_\perp^*(\boldsymbol{\kappa})\cdot\boldsymbol{P}\right\}\boldsymbol{M}_\perp(\boldsymbol{\kappa})\right.$$

$$\left.-\boldsymbol{M}_\perp^*(\boldsymbol{\kappa})\left\{\boldsymbol{M}_\perp\perp(\boldsymbol{\kappa})\cdot\boldsymbol{P}\right\}+\mathrm{i}\left\{\boldsymbol{M}_\perp^*(\boldsymbol{\kappa})\times\boldsymbol{M}_\perp(\boldsymbol{\kappa})\right\}\right], \tag{1.34}$$

where it is understood that energy is conserved;

$$E_{q'}-E_q=\frac{\hbar^2}{2m}\left(k^2-k'^2\right), \tag{1.35}$$

with k and k' the initial and final neutron wavevectors, and

$$\boldsymbol{\kappa}=\boldsymbol{k}-\boldsymbol{k}'. \tag{1.36}$$

Because $q'\neq q$ the amplitudes represented by $\boldsymbol{M}_\perp(\boldsymbol{\kappa})$ are those which connect these two states they are therefore associated with the excitation.

The cross section and final polarization expressions can be written in terms of the susceptibility giving

$$\left(\frac{d^2\sigma}{d\Omega dE}\right)_{k,k'}=\frac{V}{\pi}\frac{k'}{k}\left(\frac{e\gamma}{\hbar c}\right)^2\sum_{\alpha}\frac{\chi''_{\alpha\alpha}(\boldsymbol{\kappa},\omega)}{\left(1-e^{-\beta\hbar\omega}\right)} \tag{1.37}$$

and

$$\boldsymbol{P}'\left(\frac{\mathrm{d}^2\sigma}{\mathrm{d}\Omega\mathrm{d}E}\right)_{k,k'}=-\frac{V}{\pi}\frac{k'}{k}\left(\frac{e\gamma}{\hbar c}\right)^2\sum_{\alpha}\frac{\chi''_{\alpha\alpha}(\boldsymbol{\kappa},\omega)}{\left(1-\mathrm{e}^{-\beta\hbar\omega}\right)}\left[\boldsymbol{P}-2\mathrm{P}_{\alpha}\hat{\boldsymbol{\alpha}}\right], \tag{1.38}$$

with the sum over the two Cartesian directions perpendicular to the scattering vector.

In writing the last two equations we have selected the response to a field periodic in space and time applied in the α direction. If we had further restricted our attention to one of the two spatially helical components of this field with a helical induced response, the polarization producing last term of eqn (1.35) would have been included. This is relevant for excitations like ferromagnetic magnons where the fluctuation is helical but will not be further pursued here. Terms involving non diagonal terms of the susceptibility tensor $\chi_{\alpha\beta}$ have also not been included, but if eqn (1.21) is generalized to include nondiagonal terms then the polarization effects become even richer (Blume 1976).

1.6 MAGNETIC QUASIELASTIC SCATTERING

For small small energy transfers the cross section may be written in terms of the real part of the susceptibility. The generalized susceptibility may be written

$$\chi(\boldsymbol{\kappa},\omega)=\int_0^\infty \mathrm{d}t\,\mathrm{X}(\boldsymbol{\kappa},t)\mathrm{e}^{i\omega t}, \tag{1.39}$$

in which $\mathrm{X}(\kappa, t)$ is defined only for $t>0$. Then

$$\chi'(\boldsymbol{\kappa},\omega)=\int_0^\infty \mathrm{d}t\;\mathrm{X}(\boldsymbol{\kappa},t)\cos\omega t,$$

$$\chi''(\boldsymbol{\kappa},\omega)=\int_0^\infty \mathrm{d}t\;\mathrm{X}(\boldsymbol{\kappa},t)\sin\omega t \tag{1.40}$$

and

$$\int_{-\infty}^{\infty}\mathrm{d}\omega\frac{\chi''(\boldsymbol{\kappa},\omega)}{\omega}=\int_0^\infty \mathrm{d}t\,\mathrm{X}(\boldsymbol{\kappa},t)\int_{-\infty}^{\infty}\mathrm{d}\omega\frac{\sin\omega t}{\omega}$$

$$=\pi\chi'(\boldsymbol{\kappa},\omega=0) \tag{1.41}$$

which is a special case of the Kramers–Krönig relationships. If we now integrate the cross section over energy we have

$$\frac{\mathrm{d}\sigma}{\mathrm{d}\Omega} = \int_{-\infty}^{\infty} \hbar d\omega \left(\frac{\mathrm{d}^2\sigma}{\mathrm{d}\Omega \mathrm{d}E'} \right)$$

$$\approx \int_{-\infty}^{\infty} \hbar d\omega \left(\frac{e\gamma}{\hbar c} \right)^2 \frac{V}{\pi \beta \hbar \omega} \sum_{\alpha} \chi''_{\alpha\alpha}(\boldsymbol{\kappa}, \omega)$$

for $\hbar\omega << \frac{1}{\beta}$ and $k' \approx k$

$$= V \left(\frac{e\gamma}{\hbar c} \right)^2 k_B T \sum_{\alpha} \chi'_{\alpha\alpha}(\boldsymbol{\kappa}, 0) \tag{1.42}$$

using the Kramers–Krönig relationship derived above. Similarly

$$\boldsymbol{P}' \left(\frac{\mathrm{d}\sigma}{\mathrm{d}\Omega} \right) = -V \left(\frac{e\gamma}{\hbar c} \right)^2 k_B T \sum_{\alpha} \chi'(\boldsymbol{\kappa}, 0) \left[\boldsymbol{P} - 2P_{\alpha} \hat{\boldsymbol{\alpha}} \right]. \tag{1.43}$$

This is the quasielastic approximation for the magnetic cross section and is useful for comparison of the neutron cross section with bulk susceptibility measurements. The polarization effects are in some cases necessary for the isolation of the magnetic scattering from other sources. In particular for the polarization along the scattering vector, $P_{\alpha} = 0$, all the scattering is with neutron polarization reversal. This is a powerful signature of magnetic scattering which can only be confused with that part of the nuclear spin incoherent cross section which reverses the polarization. The two can however be experimentally distinguished by changing the direction of initial polarization which affects only the magnetic scattering final polarization.

The purpose of this chapter has been to expose and canvass the interaction of neutrons with magnetic materials. The most important point to be made is that because the neutron is a magnetic probe the cross sections measured are directly related to magnetic quantities such as the magnetization and the susceptibility. Secondly, because the neutron has a wavelength, spatial information is automatically built into the distribution of scattering via Fourier transforms of the magnetization and susceptibility. In the case of disordered materials this means that not only can the distribution of magnetic moment values be obtained but also their spatial correlation. Thirdly, through the use of polarized beam methods, the spatial atomic and magnetic distributions can be obtained separately in a wide variety of magnetic material types.

APPENDIX: Neutron Spin Transitions

The cross sections for transitions between spin states were first derived by Blume (1963) using the density matrix method. This method was also used by

Lovesey (1984*b*) and is the standard and elegant approach. It is however also possible to derive the spin dependent cross sections using the spin state functions themselves in the manner of the Schrödinger representation of quantum mechanics. For variety, and to make the arguments accessible to a wider audience it is the latter method which is outlined here.

At this point we need to consider the description of a polarized beam. The spin state of a neutron beam may be written

$$|s\rangle = \alpha|\uparrow\rangle + \beta|\downarrow\rangle . \tag{A.1}$$

The expectation values of the components of the spin are

$$\begin{aligned} \langle s|\hat{S}_z|s\rangle &= \tfrac{1}{2}\hbar(\alpha^*\alpha - \beta^*\beta) \\ \langle s|\hat{S}_x|s\rangle &= \tfrac{1}{2}\hbar(\alpha^*\alpha + \beta^*\alpha) \\ \langle s|\hat{S}_y|s\rangle &= \tfrac{1}{2}i\hbar(\beta^*\alpha - \alpha^*\beta). \end{aligned} \tag{A.2}$$

Using the Cartesian components of the Pauli spin matrices

$$\hat{\sigma}_z = \begin{pmatrix} 1 & 0 \\ 0 & -1 \end{pmatrix}; \quad \hat{\sigma}_x = \begin{pmatrix} 0 & 1 \\ 1 & 0 \end{pmatrix}; \quad \hat{\sigma}_y = \begin{pmatrix} 0 & -i \\ i & 0 \end{pmatrix} \tag{A.3}$$

these expectation values can be written

$$\begin{aligned} \langle S_z\rangle &= \tfrac{1}{2}\hbar \cos\theta \\ \langle S_x\rangle &= \tfrac{1}{2}\hbar \sin\theta\cos\phi \\ \langle S_y\rangle &= \tfrac{1}{2}\hbar \sin\theta\sin\phi, \end{aligned} \tag{A.4}$$

with the identification

$$\cos\frac{\theta}{2} = |\alpha|; \quad e^{i\phi} = \frac{\alpha|\beta|}{|\alpha|\beta}, \tag{A.5}$$

so that the spin 1/2 state function can be related to the classical motion of a spin described by the polar coordinates θ and ϕ (Mezei 1980).

Because of the interference between the two spin states of the neutron it is necessary to consider the transition probabilities rather than the transition amplitudes. Firstly, we resolve the the vectors into their components to use the properties of the Pauli matrices. Secondly, we look at the transition from a completely polarized state $|1, 0>$ to another state $|\alpha, \beta>$. The transition amplitude is

$$\langle \alpha^*,\beta^*|\langle q'|b(\boldsymbol{\kappa})|q\rangle + \langle q'|\hat{A}_z(\boldsymbol{\kappa})|q\rangle\hat{\sigma}_z + \langle q'|\hat{A}_x|q\rangle\hat{\sigma}_x + \langle q'|\hat{A}_y|q\rangle\hat{\sigma}_y|1,0\rangle . \tag{A.6}$$

It is only the neutron spin operators which act upon the neutron spin state, and only the scattering length density and magnetization and nuclear spin operators which act upon the target states. The amplitude then becomes

$$\langle q'|b(\boldsymbol{\kappa})\alpha^* + \hat{A}_z(\boldsymbol{\kappa})\alpha^* + \hat{A}_x(\boldsymbol{\kappa})\beta^* + \hat{A}_y(\boldsymbol{\kappa})\left(\mathrm{i}\beta^*\right)|q\rangle. \tag{A.7}$$

The final polarization, in general will have components along the three Cartesian directions. Appropriate spin functions for the components of polarization are:

$+z$ direction	$\alpha = 1, \beta = 0$
$-z$ direction	$\alpha = 0, \beta = 1$
$+x$ direction	$\alpha = 1/\surd 2, \beta = 1/\surd 2$
$-x$ direction	$\alpha = 1/\surd 2, \beta = -1/\surd 2$
$+y$ direction	$\alpha = 1/\surd 2, \beta = \mathrm{i}/\surd 2$
$-y$ direction	$\alpha = 1/\surd 2, \beta = \mathrm{i}/\surd 2.$

Transition amplitudes divided by $2\pi\hbar^2/m$ to the various spin states are

$$\begin{aligned}
U_z &= \langle q'|\hat{b}(\boldsymbol{\kappa}) + \hat{A}_z(\boldsymbol{\kappa})|q\rangle \\
U_{-z} &= \langle q'|\hat{A}_x(\boldsymbol{\kappa}) + \mathrm{i}\hat{A}_y(\boldsymbol{\kappa})|q\rangle \\
U_x &= \frac{1}{\sqrt{2}}\langle q'|\hat{b}(\boldsymbol{\kappa}) + \hat{A}_z(\boldsymbol{\kappa}) + \hat{A}_x(\boldsymbol{\kappa}) + \mathrm{i}\hat{A}_y(\boldsymbol{\kappa})|q\rangle \\
U_{-x} &= \frac{1}{\sqrt{2}}\langle q'|\hat{b}(\boldsymbol{\kappa}) + \hat{A}_z(\boldsymbol{\kappa}) - \hat{A}_x(\boldsymbol{\kappa}) - \mathrm{i}\hat{A}_y(\boldsymbol{\kappa})|q\rangle \\
U_y &= \frac{1}{\sqrt{2}}\langle q'|\hat{b}(\boldsymbol{\kappa}) + \hat{A}_z(\boldsymbol{\kappa}) - \mathrm{i}\hat{A}_x(\boldsymbol{\kappa}) + \hat{A}_y(\boldsymbol{\kappa})|q\rangle \\
U_{-y} &= \frac{1}{\sqrt{2}}\langle q'|\hat{b}(\boldsymbol{\kappa}) + \hat{A}_z(\boldsymbol{\kappa}) + \mathrm{i}\hat{A}_x(\boldsymbol{\kappa}) - \hat{A}_y(\boldsymbol{\kappa})|q\rangle,
\end{aligned} \tag{A.8}$$

and the equivalent transition probabilities divided by $(2\pi\hbar^2/m)^2$ for scattering with the various polarizations are

$$\begin{aligned}
\mathrm{T}'_z &= b^*(\boldsymbol{\kappa})b(\boldsymbol{\kappa}) + A_z^*(\boldsymbol{\kappa})A_z(\boldsymbol{\kappa}) + b^*(\boldsymbol{\kappa})A_z(\boldsymbol{\kappa}) + A_z^*(\boldsymbol{\kappa})b(\boldsymbol{\kappa}) \\
\mathrm{T}'_{-z} &= A_x^*(\boldsymbol{\kappa})A_x(\boldsymbol{\kappa}) + A_y^*(\boldsymbol{\kappa})A_y(\boldsymbol{\kappa}) + \mathrm{i}\{A_x^*(\boldsymbol{\kappa})A_y(\boldsymbol{\kappa}) - A_y^*(\boldsymbol{\kappa})A_x(\boldsymbol{\kappa})\} \\
\mathrm{T}'_x &= \tfrac{1}{2}[b^*(\boldsymbol{\kappa})b(\boldsymbol{\kappa}) + A_z^*(\boldsymbol{\kappa})A_z(\boldsymbol{\kappa}) + A_x^*(\boldsymbol{\kappa})A_x(\boldsymbol{\kappa}) + A_y^*(\boldsymbol{\kappa})A_y(\boldsymbol{\kappa}) \\
&\quad + b^*(\boldsymbol{\kappa})A_z(\boldsymbol{\kappa}) + A_z^*(\boldsymbol{\kappa})b(\boldsymbol{\kappa}) + b^*(\boldsymbol{\kappa})A_x(\boldsymbol{\kappa}) + A_x^*(\boldsymbol{\kappa})b(\boldsymbol{\kappa}) + \mathrm{i}\{b^*(\boldsymbol{\kappa})A_y(\boldsymbol{\kappa})
\end{aligned}$$

$$-A_y^*(\boldsymbol{\kappa})b(\boldsymbol{\kappa})\} + A_z^*(\kappa)A_x(\boldsymbol{\kappa}) + A_x^*(\boldsymbol{\kappa})A_z(\boldsymbol{\kappa}) + \mathrm{i}\{A_z^*(\boldsymbol{\kappa})A_y(\boldsymbol{\kappa})$$

$$-A_y^*(\boldsymbol{\kappa})A_z(\boldsymbol{\kappa})\} + \mathrm{i}\{A_x^*(\boldsymbol{\kappa})A_y(\boldsymbol{\kappa}) - A_y^*(\boldsymbol{\kappa})A_x(\boldsymbol{\kappa})\}]$$

$$\mathrm{T}'_{-x} = \tfrac{1}{2}\big[b^*(\boldsymbol{\kappa})b(\boldsymbol{\kappa}) + A_z^*(\boldsymbol{\kappa})A_z(\boldsymbol{\kappa}) + A_x^*(\boldsymbol{\kappa})A_x(\boldsymbol{\kappa}) + A_y^*(\boldsymbol{\kappa})A_y(\boldsymbol{\kappa}) \qquad \text{(A.9)}$$

$$+b^*(\boldsymbol{\kappa})A_z(\boldsymbol{\kappa}) + A_z^*(\boldsymbol{\kappa})b(\boldsymbol{\kappa}) - b^*(\boldsymbol{\kappa})A_x(\boldsymbol{\kappa}) - A_x^*(\boldsymbol{\kappa})b(\boldsymbol{\kappa}) - \mathrm{i}\{b^*(\boldsymbol{\kappa})A_y(\boldsymbol{\kappa})$$

$$-A_y^*(\boldsymbol{\kappa})b(\boldsymbol{\kappa})\} - A_z^*(\boldsymbol{\kappa})A_x(\boldsymbol{\kappa}) - A_x^*(\boldsymbol{\kappa})A_z(\boldsymbol{\kappa}) - \mathrm{i}\{A_z^*(\boldsymbol{\kappa})A_y(\boldsymbol{\kappa})$$

$$-A_y^*(\boldsymbol{\kappa})A_z(\boldsymbol{\kappa})\} + \mathrm{i}\{A_x^*(\boldsymbol{\kappa})A_y(\boldsymbol{\kappa}) - A_y^*(\boldsymbol{\kappa})A_x(\boldsymbol{\kappa})\}]$$

$$\mathrm{T}'_y = \tfrac{1}{2}\big[b^*(\boldsymbol{\kappa})b(\boldsymbol{\kappa}) + A_z^*(\boldsymbol{\kappa})A_z(\boldsymbol{\kappa}) + A_x^*(\boldsymbol{\kappa})A_x(\boldsymbol{\kappa}) + A_y^*(\boldsymbol{\kappa})A_y(\boldsymbol{\kappa})$$

$$+b^*(\boldsymbol{\kappa})A_z(\boldsymbol{\kappa}) + A_z^*(\boldsymbol{\kappa})b(\boldsymbol{\kappa}) - \mathrm{i}\{b^*(\boldsymbol{\kappa})A_x(\boldsymbol{\kappa}) - A_x^*(\boldsymbol{\kappa})b(\boldsymbol{\kappa})\}$$

$$+b^*(\boldsymbol{\kappa})A_y(\boldsymbol{\kappa}) + A_y^*(\boldsymbol{\kappa})b(\boldsymbol{\kappa}) - \mathrm{i}\{A_z^*(\boldsymbol{\kappa})A_x(\boldsymbol{\kappa}) - A_x^*(\boldsymbol{\kappa})A_z(\boldsymbol{\kappa})\}$$

$$+A_z^*(\boldsymbol{\kappa})A_y(\boldsymbol{\kappa}) + A_y^*(\boldsymbol{\kappa})A_z(\boldsymbol{\kappa}) + \mathrm{i}\{A_x^*(\boldsymbol{\kappa})A_y(\boldsymbol{\kappa}) - A_y^*(\boldsymbol{\kappa})A_x(\boldsymbol{\kappa})\}]$$

$$\mathrm{T}'_{-y} = \tfrac{1}{2}\big[b^*(\boldsymbol{\kappa})b(\boldsymbol{\kappa}) + A_z^*(\boldsymbol{\kappa})A_z(\boldsymbol{\kappa}) + A_x^*(\boldsymbol{\kappa})A_x(\boldsymbol{\kappa}) + A_y^*(\boldsymbol{\kappa})A_y(\boldsymbol{\kappa})$$

$$+b^*(\boldsymbol{\kappa})A_z(\boldsymbol{\kappa}) + A_z^*(\boldsymbol{\kappa})b(\boldsymbol{\kappa}) + \mathrm{i}\{b^*(\boldsymbol{\kappa})A_x(\boldsymbol{\kappa}) - A_x^*(\boldsymbol{\kappa})b(\boldsymbol{\kappa})\}$$

$$-b^*(\boldsymbol{\kappa})A_y(\boldsymbol{\kappa}) - A_y^*(\boldsymbol{\kappa})b(\boldsymbol{\kappa}) + \mathrm{i}\{A_z^*(\boldsymbol{\kappa})A_x(\boldsymbol{\kappa}) - A_x^*(\boldsymbol{\kappa})A_z(\boldsymbol{\kappa})\}$$

$$-A_z^*(\boldsymbol{\kappa})A_y(\boldsymbol{\kappa}) - A_y^*(\boldsymbol{\kappa})A_z(\boldsymbol{\kappa}) + \mathrm{i}\{A_x^*(\boldsymbol{\kappa})A_y(\boldsymbol{\kappa}) - A_y^*(\boldsymbol{\kappa})A_x(\boldsymbol{\kappa})\}],$$

where for instance $b^*(\boldsymbol{\kappa})$ has been written for $\langle q|\hat{b}^*(\boldsymbol{\kappa})|q'\rangle$.
With no analysis for polarization along any axis the transition probability is

$$\mathrm{T}_+ = \left(\frac{2\pi\hbar^2}{m}\right)^2 \big[b^*(\boldsymbol{\kappa})b(\boldsymbol{\kappa}) + A_z^*(\boldsymbol{\kappa})A_z(\boldsymbol{\kappa}) + A_x^*(\boldsymbol{\kappa})A_x(\boldsymbol{\kappa}) + A_y^*(\boldsymbol{\kappa})A_y(\boldsymbol{\kappa})$$

$$+b^*(\boldsymbol{\kappa})A_z(\boldsymbol{\kappa}) + A_z^*(\boldsymbol{\kappa})b(\boldsymbol{\kappa}) + \mathrm{i}\{A_x^*(\boldsymbol{\kappa})A_y(\boldsymbol{\kappa}) - A_y^*(\boldsymbol{\kappa})A_x(\boldsymbol{\kappa})\}]. \qquad \text{(A.10)}$$

Note that the total transition probability is that for each analysis direction independently, as analysis is done in *one* of the three Cartesian directions. With polarization reversed

$$\mathrm{T}_- = \left(\frac{2\pi\hbar^2}{m}\right)^2 \big[b^*(\boldsymbol{\kappa})b(\boldsymbol{\kappa}) + A_z^*(\boldsymbol{\kappa})A_z(\boldsymbol{\kappa}) + A_x^*(\boldsymbol{\kappa})A_x(\boldsymbol{\kappa}) + A_y^*(\boldsymbol{\kappa})A_y(\boldsymbol{\kappa})$$

$$-b^*(\boldsymbol{\kappa})A_z(\boldsymbol{\kappa}) - A_z^*(\boldsymbol{\kappa})b(\boldsymbol{\kappa}) - \mathrm{i}\{A_x^*(\boldsymbol{\kappa})A_y(\boldsymbol{\kappa}) - A_y^*(\boldsymbol{\kappa})A_x(\boldsymbol{\kappa})\}], \qquad \text{(A.11)}$$

so that if the the incident beam consists of an incoherent mixture of up and down polarizations with net polarization

$$\boldsymbol{P} = \frac{\mathrm{I}_+ - \mathrm{I}_-}{\mathrm{I}_+ + \mathrm{I}_-} \tag{A.12}$$

we have that

$$\mathrm{T} = \left(\frac{2\pi\hbar^2}{m}\right)^2 \Big[b^*(\boldsymbol{\kappa})b(\boldsymbol{\kappa}) + A^*(\boldsymbol{\kappa})\cdot A(\boldsymbol{\kappa}) + b^*(\boldsymbol{\kappa})A(\boldsymbol{\kappa})\cdot\boldsymbol{P}$$

$$+A^*(\boldsymbol{\kappa})\cdot\boldsymbol{P}\, b(\boldsymbol{\kappa}) + \mathrm{i}\left\{A^*(\boldsymbol{\kappa})\times A(\boldsymbol{\kappa})\right\}\cdot\boldsymbol{P}\Big]. \tag{A.13}$$

To calculate the final polarization $\boldsymbol{P}'$ we need to take the vector sum of the net transition probabilities along each Cartesian direction divided by the total transition probability T. In a similar manner to that used for T we obtain

$$\boldsymbol{P}'\mathrm{T} = \left(\frac{2\pi\hbar^2}{m}\right)^2 \Big[b^*(\boldsymbol{\kappa})b(\boldsymbol{\kappa})\boldsymbol{P} + b^*(\boldsymbol{\kappa})A(\boldsymbol{\kappa}) + A^*(\boldsymbol{\kappa})b(\boldsymbol{\kappa}) + \mathrm{i}b^*(\boldsymbol{\kappa})\{A(\boldsymbol{\kappa})$$

$$\times\boldsymbol{P}\} + \mathrm{i}\{\boldsymbol{P}\times A^*(\boldsymbol{\kappa})\}b(\boldsymbol{\kappa}) - \{A^*(\boldsymbol{\kappa})\cdot A(\boldsymbol{\kappa})\}\boldsymbol{P} + \{A^*(\boldsymbol{\kappa})\cdot\boldsymbol{P}\}A(\boldsymbol{\kappa})$$

$$+A^*(\boldsymbol{\kappa})\{A(\boldsymbol{\kappa})\cdot\boldsymbol{P}\} - \mathrm{i}\{A^*(\boldsymbol{\kappa})\times A(\boldsymbol{\kappa})\}\Big] \tag{A.14}$$

where it can be seen that in addition to terms which depend on an initial polarization $\boldsymbol{P}$ there are terms which require no polarization of the incident beam for a final polarization. These are the cross terms between nuclear scattering amplitude, and vector magnetic and nuclear amplitudes, and that arising from a purely vector amplitude associated with a definite direction such as a magnetic spiral structure. These are the terms used to produce polarized beams by reflection or scattering from magnetic materials.

References

Blume, M. (1963). *Phys. Rev.* **130**, 1670.

Blume, M. (1976). *Proceedings of the conference on neutron scattering*, Gatlinburg, TN, June 1976, p. 807.

Lovesey, S.W. (1984*a*). *Theory of neutron scattering from condensed matter.* Vol.1. Oxford University Press.

Lovesey, S.W. (1984b). *Theory of neutron scattering from condensed matter.* Vol.2. Oxford University Press.

Mezei, F. (1980). In *Imaging processes and coherence in physics*. Lecture notes in physics 112, p. 282. Springer, Heidelberg.

2

Dilute ferromagnetic alloys

Dilute ferromagnetic alloys are an important class of disordered magnetic materials. Because of the dilution, simple models may be employed when discussing the spatial variation of moment. The most important simplifying feature is the independence of the impurity centres so that the description of the scattering can be interpreted in terms of the effect of a single defect on the magnetism of the host. Thus the data from dilute ferromagnetic alloys is a rich source of information about the electronic state of the impurity atoms and their affect on the electronic state of the surrounding host and its spin polarization. However, before describing models particularly for dilute ferromagnetic alloys, we outline some general equations for the neutron scattering, in various situations of experimental importance, from ferromagnetic alloys in general and take a look at a classic diffractometer built especially for the small diffuse cross sections expected from dilute alloys.

2.1 DEFECT CROSS SECTIONS FOR FERROMAGNETIC ALLOYS

The purely magnetic amplitude for ferromagnets, the thermally averaged expectation value $\boldsymbol{M}(\boldsymbol{\kappa})$, takes a particularly simple form

$$\boldsymbol{M}_{\perp}(\boldsymbol{\kappa}) = \sum_i f_i(\boldsymbol{\kappa})\boldsymbol{\mu}_{\perp}(\boldsymbol{R}_i)\mathrm{e}^{\mathrm{i}\boldsymbol{\kappa}\cdot\boldsymbol{R}_i}, \tag{2.1}$$

where $f_i(\boldsymbol{\kappa})$ is the atomic magnetic form factor and $\boldsymbol{\mu}(\boldsymbol{R}_i)$ is the atomic moment at site i. In an alloy the moment at each site depends largely on the atomic species at the site as does the magnetic form factor. At this stage we make the simplification that the form factor is the same for each species. This is an approximation which is accurate enough for alloys of the transition series where the magnetic size of the atoms is similar and the diffuse scattering is dominated by variations which extend over many atoms.

The magnetic amplitude can be rewritten in the form of the average atomic moment and the deviations at each site from the average

$$\boldsymbol{M}_{\perp}(\boldsymbol{\kappa}) = f(\boldsymbol{\kappa})\sum_i \left\{\overline{\boldsymbol{\mu}}_{\perp} + \left(\boldsymbol{\mu}_{\perp}(\boldsymbol{R}_i) - \overline{\boldsymbol{\mu}}_{\perp}\right)\right\}\mathrm{e}^{\mathrm{i}\boldsymbol{\kappa}\cdot\boldsymbol{R}_i}. \tag{2.2}$$

The nuclear scattering amplitude can be written similarly

$$b(\boldsymbol{\kappa}) = \sum_i \left\{\bar{b} + \left(b(\boldsymbol{R}_i) - \bar{b}\right)\right\} \mathrm{e}^{\mathrm{i}\boldsymbol{\kappa}\cdot\boldsymbol{R}_i}. \tag{2.3}$$

By assuming that the thermal fluctuations at each site are independent, an approximation accurate to $1/N$, it is the square of the magnetic amplitude which contributes to the purely magnetic cross section. This can be written

$$\boldsymbol{M}_\perp(\boldsymbol{\kappa})\boldsymbol{M}_\perp^*(\boldsymbol{\kappa}) = f^2(\boldsymbol{\kappa})\sum_{i,j}\left\{\bar{\boldsymbol{\mu}}_\perp^2 + \left(\boldsymbol{\mu}_\perp(\boldsymbol{R}_i) - \bar{\boldsymbol{\mu}}_\perp\right)\left(\boldsymbol{\mu}_\perp(\boldsymbol{R}_j) - \bar{\boldsymbol{\mu}}_\perp\right)\right\} \mathrm{e}^{\mathrm{i}\boldsymbol{\kappa}\cdot(\boldsymbol{R}_i - \boldsymbol{R}_j)}. \tag{2.4}$$

The first term is the Bragg scattering which contributes only at specific scattering vectors $\boldsymbol{\kappa} = \mathbf{G}$, the reciprocal lattice vectors, but the second term has no sharp features and appears as diffuse scattering between the Bragg peaks. Any structure in the diffuse scattering can be seen to be due to correlations between sites of the static deviations from the average moment. The magnitude of this purely magnetic term can be manipulated with a field by changing the ferromagnetic direction with respect to the scattering vector direction, and can be extinguished if the ferromagnetic direction is along the scattering vector.

The nuclear-magnetic cross term in the cross section is

$$\boldsymbol{M}_\perp(\boldsymbol{\kappa})b^*(\boldsymbol{\kappa}) = f(\boldsymbol{\kappa})\sum_{i,j}\left\{\bar{\boldsymbol{\mu}}_\perp\bar{b} + \left(\boldsymbol{\mu}_\perp(\boldsymbol{R}_i) - \bar{\boldsymbol{\mu}}_\perp\right)\left(b(\boldsymbol{R}_j) - \bar{b}\right)\right\} \mathrm{e}^{\mathrm{i}\boldsymbol{\kappa}\cdot(\boldsymbol{R}_i - \boldsymbol{R}_j)}, \tag{2.5}$$

where it can be seen again that the first term is a contribution to the Bragg scattering and the second term is diffuse. However, this time the correlation between deviations from average values is betwen nuclear scattering lengths and magnetic moments. In other words, the correlation is between the location of atomic species and values of the magnetic moment at other sites. This term can be isolated by manipulation of the neutron polarization with respect to the ferromagnetic direction.

For an unpolarized beam and a ferromagnetic material, the diffuse only purely magnetic cross section is obtained from eqn (1.14)

$$\frac{\mathrm{d}\sigma}{\mathrm{d}\Omega} = \left(\frac{e^2\gamma}{2m_e c^2}\right)^2 f^2(\boldsymbol{\kappa})\sum_{i,j}\left(\boldsymbol{\mu}_\perp(\boldsymbol{R}_i) - \bar{\boldsymbol{\mu}}_\perp\right)\left(\boldsymbol{\mu}_\perp(\boldsymbol{R}_j) - \bar{\boldsymbol{\mu}}_\perp\right) \mathrm{e}^{\mathrm{i}\boldsymbol{\kappa}\cdot(\boldsymbol{R}_i - \boldsymbol{R}_j)}, \tag{2.6}$$

where the atomic moment values have now been converted to Bohr magnetons by multiplying the prefactor by $(e\hbar/2m_e c)^2$.

If the beam is polarized the magnetic-nuclear cross term in eqn (1.14) is added, but this is the only term which is dependent on the polarization, so that the diffuse cross section which is the difference between neutron polarization along and opposite to the ferromagnetic direction is

$$\Delta\frac{\mathrm{d}\sigma}{\mathrm{d}\Omega} = 4\left(\frac{e^2\gamma}{2m_e c^2}\right) f(\boldsymbol{\kappa})\sum_{i,j}\left(b(\boldsymbol{R}_i) - \bar{b}\right)\left(\boldsymbol{\mu}_\perp(\boldsymbol{R}_j) - \bar{\boldsymbol{\mu}}_\perp\right) \mathrm{e}^{\mathrm{i}\boldsymbol{\kappa}\cdot(\boldsymbol{R}_i - \boldsymbol{R}_j)}. \tag{2.7}$$

2.2 THE GLOPPER

Much of the early work on purely ferromagnetic diffuse scattering used an instrument popularly known as the GLOPPER on the PLUTO reactor at Harwell. The first description of the instrument appears in Low and Collins (1963) and although subsequently modified several times by the addition of detectors covering a wider range of scattering vectors and by changes in methods of data collection, it remained the same in principle. The instrument operated with an incident wavelength of about 4 Å which is beyond the Bragg cutoff for simple 3d transition series alloys, and with a crude time of flight system to select neutrons which were elastically scattered. The angular and wavelength resolutions of the instrument were very relaxed to compensate for the relatively small neutron intensity obtainable from a thermal source in a reactor at wavelengths beyond 4 Å. A schematic diagram of the instrument is shown in Fig. 2.1 Removal of the short wavelength end of the reactor spectrum was achieved by passing the beam through a filter of polycrystalline beryllium (Bragg cutoff 3.95 Å) cooled with liquid nitrogen to reduce scattering of longer wavelength neutrons by phonons. The incident beam was thus the long wavelength tail of the reactor spectrum in which the intensity decreases very quickly with wavelength and which resulted in a quasimonochromatic beam of wavelength a little more than 4 Å. This beam was then chopped by a slotted disk, and scattered by the specimen into the detector. The detector was gated to count those neutrons which were elastically scattered with wavelengths greater than 4 Å as determined by the time of flight delay setting. Most of the inelastically scattered neutrons gain energy and were thus also filtered out by the time of flight system. The wavelength resolution was ~25 per cent full width at half maximum (FWHM) and this was approximately matched by the angular resolution

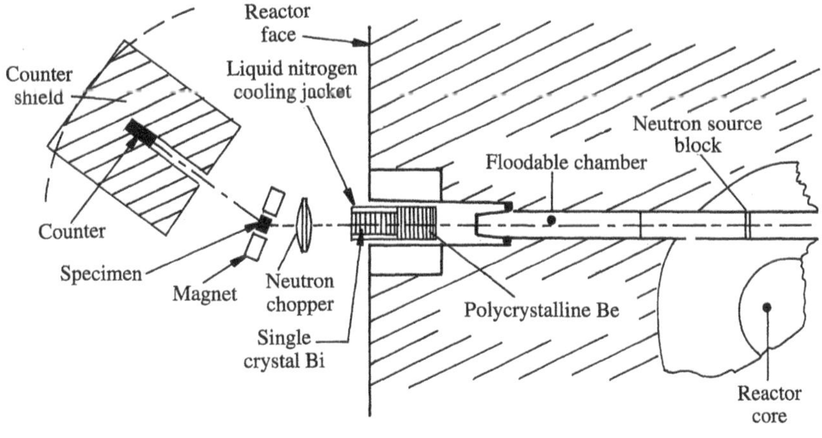

Fig. 2.1 Schematic diagram of the GLOPPER. The instrument was later modified by the addition of more detectors. (After Low and Collins 1963.)

of 2° FWHM for common scattering angles. At the specimen position there was a facility for rotating a saturating magnetic field from perpendicular to parallel to the scattering vector; from maximum to zero magnetic scattering. Alternatively, there was a facility for switching a field on and off along the scattering vector in a cryostat cooled to 4.2 K. It was assumed in the measurements that all the filtered scattering was elastic and corrections were made for beam attenuation in the sample and for multiple scattering. The latter problem was reduced by operating at long wavelengths where there is no multiple Bragg scattering which may contribute to the diffuse scattering.

2.3 DILUTE FERROMAGNETIC ALLOYS

The results from binary systems which are dilute in one constituent are particularly easy to interpret. The simplification comes about by assuming that the magnetic defect associated with each minority constituent centre is independent. This is simply what is expressed in the following equation using the operator $p_j = 1$ for species 1 at $\boldsymbol{R}_j$ and equal to zero otherwise. Then

$$\mu_j = \mu_2 + (\mu_1 - \mu_2)p_j + \sum_{i \neq j} \phi(\boldsymbol{R}_i - \boldsymbol{R}_j)p_{j+i}, \tag{2.8}$$

where the moment at j is first a function of the species of atom at $\boldsymbol{R}_j$ through p_j; and secondly a function of the presence of an atom of species 1 at $j + i$. Writing the disturbance in moment at j as the *sum* of all the effects due to species 1 at a number of sites $\boldsymbol{R}_{j+i}$, means that the magnetic defects around all impurity centres *superpose* and are thus independent. Within this model the average moment

$$\bar{\mu} = \mu_2(1-C) + \left(\mu_1 + \sum_{i \neq 0} \phi(\boldsymbol{R}_i)\right)C, \tag{2.9}$$

with C the impurity concentration, and with the recognition that for dilute alloys the average moment on the host species 2 is very close to μ_2 and that the moment on the widely separated substitutional impurities is μ_1. Because the substitution is random the cross section for any alloy will be the average over all configurations. To evaluate this average we need

$$\bar{p}_i = \bar{p}_j = C, \tag{2.10}$$

and

$$\overline{p_i p_j} = C^2 + C(1-C)\,\delta_{i,j}, \tag{2.11}$$

which are obtained by averaging over a site or pair of sites in an ensemble of alloys identical except for the distribution of impurity atoms. The pair average is done assuming no correlation between the occupation of pairs of sites. There is of course perfect correlation if the sites are the same and this is taken into account with the delta function.

Substitution of eqn (2.8) into eqn (2.6) gives

$$\frac{d\sigma}{d\Omega} = \left(\frac{e^2\gamma}{2m_e c^2}\right)^2 C(1-C) f^2(\boldsymbol{\kappa})\left\{\mu_{1_\perp} - \mu_{2_\perp} + \Phi_\perp(\boldsymbol{\kappa})\right\}^2, \tag{2.12}$$

and substitution into eqn (2.7) gives

$$\Delta\frac{d\sigma}{d\Omega} = 4\left(\frac{e^2\gamma}{2m_e c^2}\right) C(1-C)(b_1 - b_2) f(\boldsymbol{\kappa})\left\{\mu_{1_\perp} - \mu_{2_\perp} + \Phi_\perp(\boldsymbol{\kappa})\right\}. \tag{2.13}$$

The moment change per impurity atom using this model is

$$\Delta\bar{\mu} = \frac{d\bar{\mu}}{dC} = \mu_1 - \mu_2 + \Phi(0). \tag{2.14}$$

The $\boldsymbol{\kappa} = 0$ cross sections obtained from this model are thus directly related to the macroscopic change of moment with concentration. This is an example of a much more general relationship independent of any modelling of the moment distribution. This will be explored further in Chapter 3.

2.4 DATA FROM DILUTE NICKEL ALLOYS

Figure 2.2 shows data on impurities in nickel obtained by Comly *et al.* (1968). The data were obtained on the GLOPPER by subtracting the scattering with the field along the scattering vector from that obtained with field applied perpendicular to the scattering vector. The data correspond to eqn (2.12) but expressed per impurity atom, and the alloys were assumed to be random because of the low (1–4 per cent) impurity concentrations. For a wide variety of solutes which decrease the average nickel moment the $\boldsymbol{\kappa}$ dependence of the scattering is almost identical. The cross sections in Fig. 2.2 are presented including a scaling factor W^2 to demonstrate this point. The arrows on the abscissa mark the $\boldsymbol{\kappa} = 0$ cross sections expected from the macroscopic $d\bar{\mu}/dC$ measurements. Comly *et al.* (1968) interpret the magnetic defect as arising from, firstly, a virtual bound state extending no further than first neighbours of the impurity and then secondly, to a propagation of the central loss in moment to further distances by an exchange controlled process. The second mechanism dominates the low $\boldsymbol{\kappa}$ variation in cross section which, because it is a function of the nickel matrix, is independent of the solute. It is the amplitude of the defect which is then dependent on solute. Comly *et al.* (1968) make the comment that the zinc solute does not appear to completely fit the above scheme and this trend is confirmed by later data taken on copper impurities in nickel, where the defect is confined to first neighbour (Cable *et al.* 1969; Aldred *et al.* 1973). It appears, within the mechanisms assumed by Comly *et al.*(1968) that for small disturbances (for example copper) the defect shape is dominated by the bound state formation, whereas for large disturbances the defect shape is dominated by the exchange produced penetration of the defect into the nickel matrix.

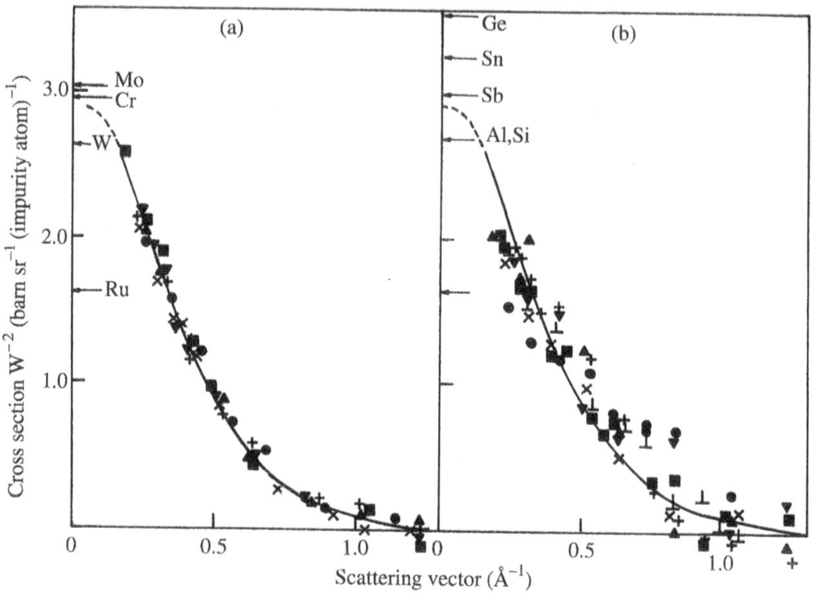

Fig. 2.2 Combined plots of the purely magnetic scattering from nickel alloys. (a) (●) Cr,0.94; (▲) Nb,0.94; (▼) Mo,0.93; (■) Ru,0.64; (+) W,1.00; (×) Re,0.99. (b) (●) Zn,0.446; (▲) Al,0.461; (▼) Ga,0.494; (■) Si,0.620; (×) Ge,0.530; (+) Sn,0.631; (⊥) Sb,0.830. The value of W used to normalize each data set follows each impurity element. (After Comly *et al.* 1968.)

Vanadium is another impurity which reduces the nickel moment and does not fit into the above scheme, but as Collins and Low (1965) point out, the difference lies in the appearance of a negative moment at the impurity site of $2 \pm 0.4\mu_B$. In this case there is also the widespread disturbance in the nickel matrix moment appropriate for large disturbances. The cross section at large $\boldsymbol{\kappa}$ was taken for the determination of the moment at the impurity site, and it was assumed that $\Phi(\boldsymbol{\kappa})\rightarrow 0$ for $\boldsymbol{\kappa}$ large. A measure of $\mu_1 - \mu_2$ is then obtained and, apart from the ambiguity of sign due to the square in eqn (2.12), μ_1 can be determined if μ_2 (the matrix moment per atom) is known. By this means it was ascertained that for all the solutes examined except vanadium the moment at the impurity site is within $1/2 \pm 1\mu_B$. More recently, by using the polarization dependent cross section, Cable and Medina (1976) and Medina and Cable (1977) have determined the moments at copper and chromium in nickel to be $0 \pm 0.01\mu_B$ and $0 \pm 0.1\mu_B$.

The other type of impurities in nickel are typified by the results on Ni–Mn and Ni–Fe shown in Fig. 2.3 along with Ni–V and Ni–Cr results. Additions of iron, manganese, and rhodium increase the moment per atom of the alloy over pure nickel by an addition of moment at the impurity site. The lack of $\boldsymbol{\kappa}$ dependence in the scattering other than that due to $|f(\boldsymbol{\kappa})|^2$ means that $\Phi(\boldsymbol{\kappa})\approx 0$ and thus values of μ_1 can be obtained from the magnitude of the scattering. The moment values

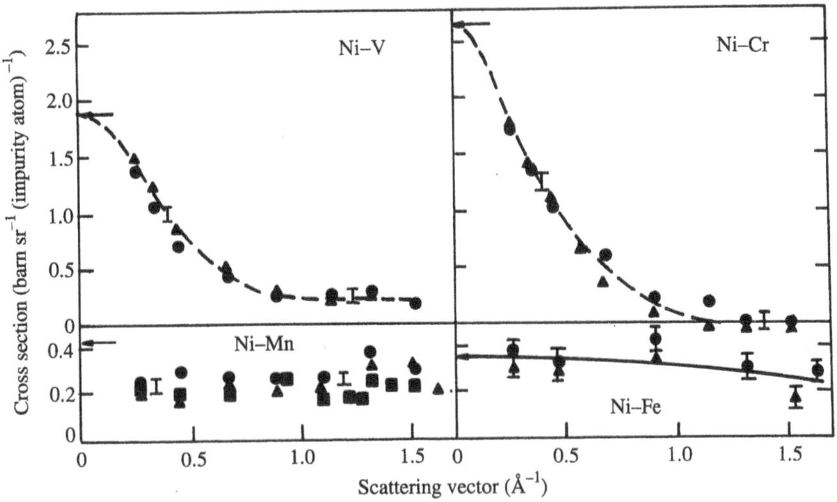

Fig. 2.3 Magnetic scattering from nickel alloys showing a range of types of variation with κ. Note that the flat distribution of scattering from Ni–Mn and Ni–Fe which indicates a magnetic defect restricted to the impurity site. (After Collins and Low 1965.)

obtained are $2.8 \pm 0.1\mu_B$ at iron sites, $2.4 \pm 0.1\mu_B$ at manganese sites, and $\sim 2.6\mu_B$ at rhodium sites. Comly *et al.* (1968) point out that because one of the spin-polarized d bands in nickel is full, it will be much more difficult to increase the moment in the nickel matrix than to reduce it. Hence impurities which raise the average moment must do so at the impurity site.

2.5 DILUTE COBALT ALLOYS

Cobalt is the least studied of the ferromagnetic 3d metals possibly because of its higher neutron absorption cross section. Another problem is that some solutes like iron depress the temperature of the hexagonal to f.c.c. structure transformation so that it may be difficult to make a single phase sample. Nevertheless, the diffuse cross sections of a few transition metal solutes in cobalt have been measured. Figure 2.4 shows measurements taken by Cable and Hicks (1970) in a series of alloys with 5 at% of solute. When first cast, the alloys contained both phases in approximately equal proportions but after hot rolling all except the Co–Mo alloy were converted to the h.c.p. phase. In the two cases where measurements were taken of the samples in both metallurgical conditions there was very little difference in the diffuse scattering indicating a moment reduction mechanism very little dependent on cobalt structure. The h.c.p. structure presents another problem for measurements of this kind which are due to the lower symmetry and high magnetic anisotropy associated with the structure. Cable and Hicks (1970) found it difficult to saturate their samples and because of possible preferred orientation induced by hot rolling they were not sure whether the

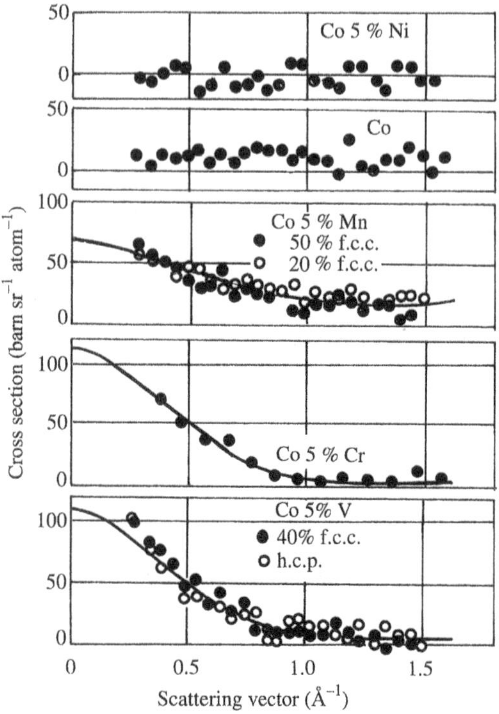

Fig. 2.4 The purely magnetic scattering from various cobalt alloys. Note the similarity to nickel alloys. (After Cable and Hicks 1970.)

directions of easy magnetization were randomly distributed in their samples. Thus, they they could not be sure that the cross section from the demagnetized sample was exactly two-thirds of the maximum expected from eqn (2.12).

The picture of moment disturbances one can obtain from these results is that vanadium and chromium reduce the moment in the cobalt matrix in a similar way to that of impurities in nickel, and that the moment at the vanadium or chromium sites is $0 \pm 0.7\mu_B$. Also manganese reduces the moment in the cobalt-matrix but not nearly so much; the moment at the manganese site is $-1 \pm 0.7\mu_B$. The diffuse scattering for nickel in cobalt was masked by the uncertainties in the experiment but it can be said that nickel produces very little reduction of moment in the cobalt matrix.

2.6 DILUTE IRON ALLOYS

The types of diffuse scattering observed for dilute impurities in iron are too diverse to cover in this section. Comprehensive sets of data for purely magnetic scattering can be found in the papers of Collins and Low (1965) and Holden *et al.*

(1967). Very little data using the polarization dependent cross section are available because of the large amount multiple Bragg scattering expected at the usual wavelength of 1.07Å. Some work, however, has been done at longer wavelengths by Kajzar and Parette for copper impurities (1979*a*), for vanadium, chromium, and manganese (1979*b*) and for titanium impurities (1979*c*) using polarized neutrons. The electronic processes which have been proposed for the magnetic distributions observed in these experiments are covered in part in the papers of Campbell (1966), Campbell and Gomès (1967) and Low (1967).

Our purpose here is to discuss one particular type of scattering observed for dilute impurities in iron as a further example of the interpretation of such scattering. Figure 2.5 shows some of the results of Collins and Low (1965) in which 4d and 5d elements have been substituted into iron. Notice the interesting scattering observed for the substitution of palladium, platinum, rhodium, and iridium. The purely magnetic cross section appears to go to zero at approximately 0.5 Å^{-1}; cobalt and nickel also exhibit this behaviour. To explain this we recall the expression for the purely magnetic cross section in eqn (2.12) which is proportional to $\{\mu_1 - \mu_2 + \Phi(\boldsymbol{\kappa})\}^2$. This will go to zero if $\{\mu_1 - \mu_2 + \Phi(\boldsymbol{\kappa})\}$changes sign as a function of $\boldsymbol{\kappa}$. Because $\Phi(\boldsymbol{\kappa})\rightarrow 0$ as $\boldsymbol{\kappa}\rightarrow$large, the implication of this is that $\Phi(0)$ is opposite in sign to $\mu_1 - \mu_2$, and larger in magnitude than $\mu_1 - \mu_2$. $\Phi(0)$ is the total moment defect other than at the impurity site, which means the sign of $d\bar{\mu}/dC$ is opposite to the change of moment at the impurity site. For all

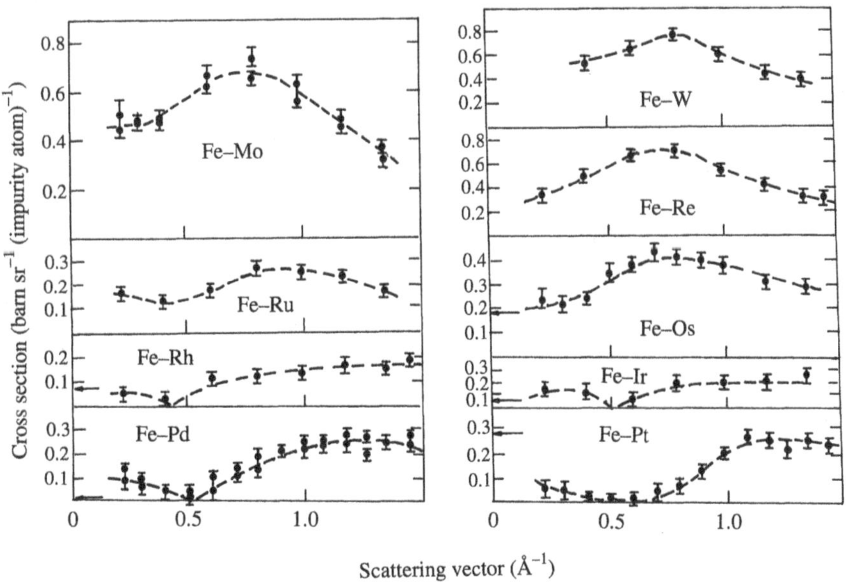

Fig. 2.5 Purely magnetic scattering from various iron alloys. Note especially the alloys for which the scattering is zero near κ ~0.5 Å^{-1}. This indicates a defect at the impurity site of opposite sign to the total defect as described in the text. (After Collins and Low 1965.)

the alloys which show this type of scattering $d\bar{\mu}/dC$ is positive so that the *defect* at the impurity site must be negative and in fact analysis of the high κ scattering gives moments at the impurity site close to zero (a defect ~ $-2.2\mu_B$).

Another interesting feature of this type of scattering is how precisely (for example Fe–Pd) the scattering from these polycrystalline alloys goes to zero at ~0.5 Å^{-1}. This means that the cross section from each domain must be closely isotropic with respect to the *direction* of $\boldsymbol{\kappa}$ otherwise averaging the cross section at $|\boldsymbol{\kappa}| = 0.5$ Å^{-1} over all $\boldsymbol{\kappa}$ directions would produce a nonzero polycrystalline cross section. Collins and Low (1965) assumed that the defect was spherically symmetric and Fourier transformed the amplitudes from the polycrystalline data to give a picture of the defect in real space. They obtained a positive moment disturbance in the matrix which decreases with distance from the impurity site. The diffuse scattering thus gives an explicit picture of a rather complicated moment defect in which the total change in moment is a competition between impurity site and matrix defects.

2.7 DILUTE IMPURITIES IN EXCHANGE ENHANCED HOSTS

One other class of dilute ferromagnetic alloy deserves mention here: alloys in which a strongly paramagnetic host is driven ferromagnetic by an impurity with a well-developed moment. The classic example is Pd–Fe. The data in Fig. 2.6 are the purely magnetic cross sections for a single crystal of palladium ~0.25 per cent iron with $\boldsymbol{\kappa}$ along the three principal directions. Hicks *et al.* (1968) in interpeting these data used a particularly simple model in which it was assumed that the moment defect was propagated into the matrix by magnetic interactions. Specifically the expression for a moment at $\boldsymbol{R}_m$ with the impurity at $\boldsymbol{R} = 0$ is written

$$\phi(\boldsymbol{R}_m) = \chi_0\left\{\sum_{n\neq m}\mathscr{I}(\boldsymbol{R}_m - \boldsymbol{R}_n)\phi(\boldsymbol{R}_n) + h_0\delta(\boldsymbol{R}_m)\right\}, \tag{2.15}$$

(in this case the moment disturbances ϕ are also the moments, as no static moment exists before the addition of impurities), where $\mathscr{I}(\boldsymbol{R}_m - \boldsymbol{R}_n)$ is a magnetic interaction parameter (like an exchange interaction), h_0 an effective field at the impurity, χ_0 the susceptibility of a matrix atom. Fourier transforming (2.15) we have

$$\Phi(\boldsymbol{\kappa}) = \chi_0\left\{I(\boldsymbol{\kappa})\Phi(\boldsymbol{\kappa}) + h_0\right\} = h_0\chi_0/(1 - \chi_0 I(\boldsymbol{\kappa})). \tag{2.16}$$

Expanding $I(\kappa)$ to second order, we have

$$\Phi(\boldsymbol{\kappa}) = h_0\chi_0/\left[1 - \chi_0 I(0)\left\{1 - g|\boldsymbol{\kappa}|^2\right\}\right]. \tag{2.17}$$

We thus have a Lorentzian form to substitute for $\Phi(\boldsymbol{\kappa})$ in eqns (2.12) and (2.13). This simple model is important in that it fits the $\boldsymbol{\kappa}$ variation of the cross sections for dilute Pd–Fe and because it predicts that $\Phi(\boldsymbol{\kappa})$ is almost independent of the

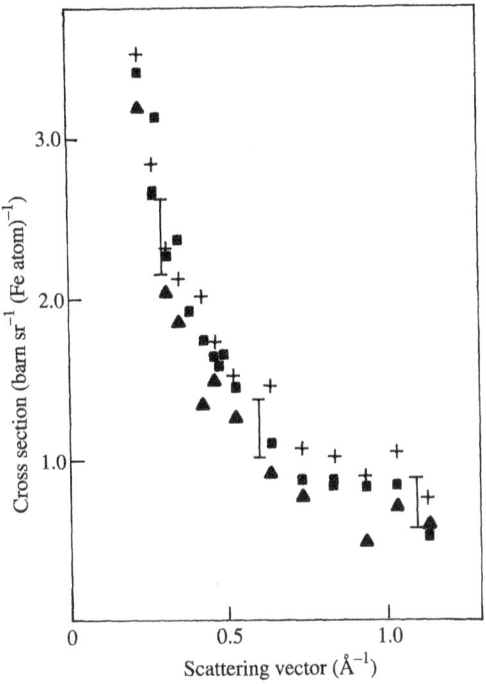

Fig. 2.6 The purely magnetic scattering from a palladium~0.25 at% iron single crystal along three principal directions: (■)< 1 0 0>; (+)< 1 1 0>; (▲)< 1 1 1>. (After Hicks *et al.* 1968.)

direction of **κ** (see Fig.2.6). Although $I(\boldsymbol{\kappa})$ for palladium must have cubic symmetry, the expansion to only second order neglects anything but the spherically symmetric part. The total moment associated with an iron impurity in palladium is giant ($10\mu_B$), whereas the the moment at the iron site is $\sim 3.5 \pm 0.4\mu_B$. The moment also has a giant radius of approximately 10 Å encompassing some 200 palladium atoms. The Pd–Co (Low and Holden 1966) and platinum base alloys give similar results. In all the giant moment alloys overlap effects become important at very low concentrations and something more sophisticated than the independent impurity model is required.

References

Aldred, A.T., Rainford, B.D., Hicks, T.J., and Kouvel, J.S. (1973). *Phys. Rev. B* **7**, 218.
Cable, J.W., Wollan, E.O., and Child, H.R. (1969). *Phys. Rev. Lett.* **22**, 1256.
Cable, J.W. and Hicks, T.J. (1970). *Phys. Rev. B* **2**, 176.
Cable, J.W. and Medina, R.A. (1976). *Phys. Rev. B* **13**, 4868.
Campbell, I.A. (1966). *Proc. Phys. Soc.* **89**, 71.
Campbell, I.A. and Gomès, A.A. (1967). *Proc. Phys. Soc.* **91**, 319.

Collins, M.F. and Low, G.G. (1965). *Proc. Phys. Soc.* **86**, 535.
Comly, J.B., Holden, T.M., and Low, G.G. (1968). *J. Phys. C* **1**, 458.
Hicks, T.J., Holden, T.M., and Low, G.G. (1968). *J. Phys. C* **1**, 528.
Holden, T.M., Comly, J.B., and Low, G.G. (1967). *Proc. Phys. Soc.* **92**, 726.
Kajzar, F. and Parette, G. (1979*a*). *J. Appl. Phys.* **50**, 1966.
Kajzar, F. and Parette, G. (1979*b*). *J. Mag. Mag. Mat.* **14**, 253.
Kajzar, F. and Parette, G. (1979*c*). *Phys. Rev. B* **20**, 2002.
Low, G.G. and Holden, T.M. (1966). *Proc. Phys. Soc.* **89**, 119.
Low, G.G. (1967). *Proc. Phys. Soc.* **92**, 938.
Low, G.G. and Collins, M.F. (1963). *J. Appl. Phys.* **34**, 1195.
Medina, R.A. and Cable, J.W. (1977). *Phys. Rev. B* **15**, 1539.

3

Concentrated ferromagnetic alloys

The analysis of defect scattering from concentrated ferromagnetic alloys is much more complicated than for dilute alloys. The effects of the defects now overlap so that they are no longer independent. There are a number of ways of dealing with this. Firstly, one can treat an alloy of one composition as a distribution of identical superposing defects. This is the early linear model of Marshall (1968). However, it is rarely the case that the defects do superpose in the sense that the addition of an extra impurity atom would add another identical defect. Normally the amplitude and extent of the defect in the moment distribution changes with composition. The defects obtained by this sort of analysis are thus an average for each particular composition. The linear model can be modified by including a correction to the effect of one impurity atom by the presence of a second (Balcar and Marshall 1968) but then the extension to triples and beyond becomes cumbersome. Some progress can be made by modelling the way in which the overlap of the effects of impurity atoms may be nonlinear. This is done in the magnetic environment model which is discussed in detail later in this chapter.

One of the most exciting recent developments has been the emergence of first principles theories of compositional and magnetic correlations in disordered metallic alloys. These are based on the self consistent field, Korringa–Kohn–Rostoker, coherent potential approximation method, and within the local spin density approximation (Staunton *et al.* 1992). The only comparison of defect scattering with this theory is that of Staunton *et al.* (1990) for compositional correlations in Fe–V alloys which is very good. No doubt detailed comparisons with the data for magnetic correlations will appear in due course, but for this chapter we will concentrate on the various phenomenonological models outlined above.

3.1 ATOMIC SHORT RANGE ORDER

The distribution of atoms in a ferromagnetic alloys has a profound effect on the distribution of moment. Even if the magnetic moments associated with the individual atomic species were not a function of their environment the distribution of moment would depend on the atomic distribution. Dependence of the moment on the environment adds another atomic distribution factor to the moment distribution.

Fortunately, neutron scattering allows the determination of the atomic distribution in the same specimen as the magnetic distribution. In fact, for concentrated alloys, neutron scattering is the best available method for the determination of atomic short range order in most cases. Using the thermally averaged nuclear amplitude $b(\boldsymbol{\kappa})$ the purely nuclear part of the elastic cross section (1.14) can be written

$$\frac{d\sigma}{d\Omega} = \sum_{i,j} \left\{ \bar{b}^2 + \left(b(\boldsymbol{R}_i) - \bar{b}\right)\left(b(\boldsymbol{R}_j) - \bar{b}\right) \right\} e^{i\boldsymbol{\kappa}\cdot(\boldsymbol{R}_i - \boldsymbol{R}_j)} . \tag{3.1}$$

The first term in the brackets is the Bragg scattering, proportional to the square of the average nuclear scattering length, and the second term is diffuse scattering. The nuclear amplitude at any site is

$$b_i = b_2 + (b_1 - b_2)p_i , \tag{3.2}$$

with $p_i = 1$ for a type 1 atom at i and zero otherwise. Because of the statistical nature of the distribution the configurational average is needed. The single site average of p is

$$\bar{p}_i = C , \tag{3.3}$$

the concentration of type 1 atoms. For concentrated alloys we need to allow for correlations between the occupation of sites in the pair average, so that

$$\overline{p_i p_j} = C^2 + C(1 - C)\alpha_{ij} , \tag{3.4}$$

with

$$\alpha_{ij} = \frac{\overline{p_i p_j} - C^2}{C(1 - C)} . \tag{3.5}$$

The Cowley short range order parameter, α_{ij}, is equal to 1 for $i = j$ because each atom is perfectly correlated with itself. However the off site Cowley short range order parameters are generally nonzero to describe correlations in the occupation of neighbouring sites.

Substitution of the Cowley short range order parameter into eqn (3.1) gives for the diffuse part of the cross section

$$\frac{d\sigma}{d\Omega} = NC(1 - C)(b_1 - b_2)^2 S(\boldsymbol{\kappa}) , \tag{3.6}$$

in which

$$S(\boldsymbol{\kappa}) = \sum_{i,j} \alpha_{ij} e^{i\boldsymbol{\kappa}\cdot(\boldsymbol{R}_i - \boldsymbol{R}_j)} . \tag{3.7}$$

$S(\boldsymbol{\kappa})$ can be used to characterize the range and extent of the correlations in atomic occupations and to factor this into models which describe the affect of atomic short range order on the magnetic moment distribution.

3.2 MACROSCOPIC LIMITS

It is important whenever possible to make a connection between measurements on an atomic scale such as these, and macroscopic measurements. In this case it is possible to express the diffuse cross sections in terms of macroscopic variables independently of any model for the moment distribution. This also reveals what factors in the atomic distribution affect the cross section. For $\boldsymbol{\kappa} = 0$ eqn (2.6) for the purely magnetic diffuse cross section is

$$\frac{d\sigma}{d\Omega} = \left(\frac{e^2\gamma}{2m_e c^2}\right)^2 \sum_{i,j} \overline{(\mu_i - \overline{\mu})(\mu_j - \overline{\mu})}$$

$$= \left(\frac{e^2\gamma}{2m_e c^2}\right)^2 \left(\overline{M^2} - \left(\overline{M}\right)^2\right), \tag{3.8}$$

in which M is the total moment of a sample and the average is over an ensemble of equivalently prepared samples and is accurate to $1/N$. But

$$\left(\overline{M^2} - \left(\overline{M}\right)^2\right) = \overline{\left(M - \overline{M}\right)^2} = \overline{\left(\Delta M\right)^2}, \tag{3.9}$$

and we can then see that spatial fluctuations of any factors which produce a change in total moment from sample to sample will also produce a spatial fluctuation in moment in each sample and thus a diffuse cross section.

Marshall (1968), who first produced this argument, identified two factors (a) fluctuations in the concentration, and (b) fluctuations in the atomic short range order of the atoms in the alloy. So for a binary alloy consisting of atoms of types 1 and 2

$$\Delta M = \frac{dM}{dN_1}\Delta N_1 + \frac{dM}{d\alpha_{12}}\Delta\alpha_{12}, \tag{3.10}$$

where N_l is the number of type 1 atoms and $N_l = NC$ with C the concentration, and α_{l2} is a parameter describing the short range order of the atoms in the alloy. The contribution arising from concentration fluctuations can be further analysed so that

$$\overline{\left(\Delta M\right)^2} = \left(\frac{d\overline{\mu}}{dC}\right)^2 \overline{\left(\Delta N_1\right)^2}, \tag{3.11}$$

using $M = N\overline{\mu}$. If we define an operator p_i to be 1 if an atom of type 1 is at i and zero otherwise we can express

$$\overline{N}_1 = \sum_i \overline{p}_i = NC, \tag{3.12}$$

and

$$\overline{N_l^2} = \sum_{i,j} \overline{P_i P_j} = N^2C^2 + NC(1-C), \tag{3.13}$$

for a random alloy. Hence

$$\overline{(\Delta N_1)^2} = \overline{N_1^{\,2}} - \left(\overline{N_1}\right)^2 = NC(1-C) \ , \tag{3.14}$$

and

$$\frac{\mathrm{d}\sigma}{\mathrm{d}\Omega} = \left(\frac{e^2\gamma}{2m_e c^2}\right)^2 C(1-C)\left(\frac{\mathrm{d}\bar{\mu}}{\mathrm{d}C}\right)^2 + \text{s.r.o. fluctuation term} \tag{3.15}$$

at a vanishingly small scattering vector. For a nonrandom alloy the average short range order (s.r.o.) may be introduced into the concentration fluctuation term by writing

$$\overline{N_l^{\,2}} = \sum_{i,j} \overline{p_i p_j} = N^2C^2 + NC(1-C)\sum_{i,j}\alpha_{ij} \ , \tag{3.16}$$

where α_{ij} are the Cowley short range order (s.r.o.) parameters. The general form for $\mathrm{d}\sigma/\mathrm{d}\Omega$ is then

$$\frac{\mathrm{d}\sigma}{\mathrm{d}\Omega} = \left(\frac{e^2\gamma}{2m_e c^2}\right)^2 C(1-C)\left(\frac{\mathrm{d}\bar{\mu}}{\mathrm{d}C}\right)^2 \sum_{i,j}\alpha_{ij} + \text{s.r.o. fluctuation term} \, . \tag{3.17}$$

Strictly $\Sigma_{ij}\alpha_{ij} = 1$ (omitting any behaviour under the Bragg peak at $\boldsymbol{\kappa} = 0$) but experiments generally do not produce data close enough to $\boldsymbol{\kappa} = 0$ for this limit to be reached and the appropriate experimental extrapolations yield mostly $\Sigma_{ij}\alpha_{ij} \neq 1$.

A limit for $\boldsymbol{\kappa} = 0$ can also be obtained for the polarization dependent part of the diffuse cross section

$$\Delta\frac{\mathrm{d}\sigma}{\mathrm{d}\Omega} = 4\left(\frac{e^2\gamma}{2m_e c^2}\right)\sum_{i,j}\overline{(b_i - \bar{b})(\mu_j - \bar{\mu})} \, , \tag{3.18}$$

which in a manner analogous to eqn (3.9) reduces to

$$\Delta\frac{\mathrm{d}\sigma}{\mathrm{d}\Omega} = 4\left(\frac{e^2\gamma}{2m_e c^2}\right)\overline{\Delta B_t \Delta M} \, , \tag{3.19}$$

where $B_t = \Sigma_i b_i$ is the total scattering length and

$$\Delta B_t = B_t - \overline{B}_t = \frac{\mathrm{d}B}{\mathrm{d}N_1}\Delta N_1 = (b_1 - b_2)\Delta N_1 \, . \tag{3.20}$$

This gives

$$\begin{aligned}\Delta\frac{\mathrm{d}\sigma}{\mathrm{d}\Omega} &= 4\left(\frac{e^2\gamma}{2m_e c^2}\right)(b_1 - b_2)\frac{\mathrm{d}\bar{\mu}}{\mathrm{d}C}\overline{(\Delta N_1)^2} \\ &= 4\left(\frac{e^2\gamma}{2m_e c^2}\right)NC(1-C)(b_1 - b_2)\frac{\mathrm{d}\bar{\mu}}{\mathrm{d}C}\sum_{i,j}\alpha_{ij} \, , \end{aligned} \tag{3.21}$$

for the contribution due to concentration fluctuations. We can also show that there is no contribution in this cross section due to fluctuations in short range order. This contribution to $d\sigma/d\Omega$ can be written as proportional to $(dM/d\alpha_{12})(b_1 - b_2)\overline{\Delta N_1 \Delta \alpha_{12}}$, but as N_1 and α_{12} are independent variables

$$\overline{\Delta N_1 \Delta \alpha_{12}} = \overline{\Delta N_1}\ \overline{\Delta \alpha_{12}} = 0\,, \tag{3.22}$$

so that there is no contribution at $\boldsymbol{\kappa} = 0$ to the polarization dependent part of the diffuse cross section arising from fluctuations of short range order [$\boldsymbol{\kappa} = 0$ limits are discussed more generally by Medina and Garland (1976)]. There are thus two different parts of the diffuse cross section which yield slightly different information about the moment distribution in a binary ferromagnetic alloy. At $\boldsymbol{\kappa} = 0$ the general relations derived in terms of bulk variables show different dependences in that the purely magnetic cross section contains information about fluctuations in short range order in addition to those due to fluctuations in concentration. The polarization dependent cross section shows only the dependence on concentration fluctuations.

3.3 MODELS FOR CONCENTRATED FERROMAGNETIC ALLOYS

A wealth of data exists in diffuse scattering measurements on concentrated binary ferromagnetic alloys. Because of the $C(1 - C)$ term, cross sections in concentrated alloys have been easier to measure, but because of the inapplicability in most case of the assumption of independent defects the results have been harder to interpret.

The first measurements of the purely magnetic cross sections were made by Shull and Wilkinson (1955) on disordered alloys of Ni–Fe, Fe–Cr, and Co–Cr. In their interpretation they assumed moments which depend only on their atomic species and not on the surroundings. In this way they could take advantage of the independent defect model right across the composition range. The analysis is then exactly analogous to that of the diffuse nuclear neutron scattering (or also diffuse X-ray scattering) where the scattering amplitude at each site is only a function of the atom species at that site. This we know to be exactly true for nuclear neutron scattering because the scattering amplitude is a nuclear property. It is less true but still very accurate for diffuse X-ray scattering from binary alloys because only a small part of the X-ray scattering amplitude comes from electrons which are shared by more than one atom, and are thus not characteristic of any one atom. In treating diffuse magnetic scattering from binary 3d metal alloys we must be very wary of the independent atom assumption because most of the scattering comes from those electrons which are widely shared. Nevertheless, there appear to be a number of ferromagnetic alloy systems in which it is approximately true, and this is evident for instance in the data of Shull and Wilkinson (1955) on Ni–Fe and Co–Fe where the $\boldsymbol{\kappa}$ dependences of the purely magnetic and the nuclear defect scattering are closely similar. This indicates that the $\boldsymbol{\kappa}$ dependence in the magnetic scattering is simply a reflection of correlations in the placement of the atoms, and this is confirmed by the consistent changes of both nuclear and magnetic cross sections with the rearrangement of atoms by heat treatment. Even in these systems however, the change of average moments on the two species with composition means that the moments are to some extent a function of their environment. For other alloy systems it is essential to model the effect of the environment on the moments.

The first real attempt to go beyond the independent atom approach for concentrated alloys was by Marshall (1968). In the same paper in which he outlined the general $\boldsymbol{\kappa} = 0$ limit which we have reproduced in eqns (3.15) and (3.17), he suggested a model in which the moment depends on the local *atomic* environment. Here we discuss Marshall's model with only one magnetic species in an effort to simplify the algebra and expose the principle. If we suppose that species 2 carries a moment, then Marshall's model can be written

$$\mu_j = (1 - p_j)\left\{\bar{\mu}_2 + \sum_{j \neq i} \phi(\boldsymbol{R}_j - \boldsymbol{R}_i)(p_i - C)\right\}, \tag{3.23}$$

which with some algebra gives

$$\frac{\mathrm{d}\sigma}{\mathrm{d}\Omega} = \left(\frac{e^2\gamma}{2m_e c^2}\right)^2 \Big\{C(1-C)\left[\bar{\mu}_2 - (1-C)\Phi(\boldsymbol{\kappa})\right]^2 +$$

$$C^2(1-C)^2 \sum_{j \neq i} \phi^2(\boldsymbol{R}_j - \boldsymbol{R}_i)\left[1 + \exp\left(\mathrm{i}\boldsymbol{\kappa} \cdot (\boldsymbol{R}_j - \boldsymbol{R}_i)\right)\right]\Big\} \tag{3.24}$$

for a random alloy, with the $\phi(\boldsymbol{R}_j - \boldsymbol{R}_i)$ the moment perturbation at $\boldsymbol{R}_j$ due to the presence of an atom of type 1 at $\boldsymbol{R}_i$.

Firstly we note

$$\frac{\mathrm{d}\bar{\mu}}{\mathrm{d}C} = -\bar{\mu}_2 + (1-C)\frac{\mathrm{d}\bar{\mu}_2}{\mathrm{d}C}, \tag{3.25}$$

and secondly we note from eqn (3.23) that

$$\frac{\mathrm{d}\bar{\mu}_2}{\mathrm{d}C} = \Phi(0)\ . \tag{3.26}$$

We can then confirm

$$\frac{\mathrm{d}\sigma}{\mathrm{d}\Omega} = \left(\frac{e^2\gamma}{2m_e c^2}\right)^2 \left\{C(1-C)\left(\frac{\mathrm{d}\bar{\mu}}{\mathrm{d}C}\right)^2 + \text{const} \times C^2(1-C)^2\right\}, \tag{3.27}$$

which is the limiting $\boldsymbol{\kappa} = 0$ form derived in eqns (3.15) and (3.17) if we identify the second term with fluctuations in $\bar{\mu}$ caused by fluctuations in short range order (s.r.o.). We can argue that for this model such a term will arise by considering the redistribution of type 2 atoms from type 2 poor regions (a change in atom distribution and therefore in s.r.o.). With such a redistribution $\bar{\mu}$ will increase because the presence of extra type 2 atoms in type 2 rich regions will positively perturb the moments on the many type 2 atoms in the new environment, more than compensating for the loss of moment in type 2 poor regions because of the fewer type 2 atoms there. Thus there is a correlation between fluctuations of total moment and atomic short range order leading to a term in $d\sigma/d\Omega$ beyond the concentration fluctuation term. For this model and a random alloy

$$\Delta\frac{d\sigma}{d\Omega}=\left(\frac{e^2\gamma}{2m_ec^2}\right)C(1-C)(b_2-b_1)\left[\bar{\mu}_2-(1-C)\Phi(\boldsymbol{\kappa})\right], \tag{3.28}$$

so that

$$\Delta\frac{d\sigma}{d\Omega}=\left(\frac{e^2\gamma}{2m_ec^2}\right)C(1-C)(b_1-b_2)\frac{d\bar{\mu}}{dC}, \tag{3.29}$$

again in accord with the general equations (3.18), (3.21) and specifically with no further terms due to fluctuation in s.r.o. The average moment for such an alloy system is obtained by integrating eqn (3.25) to yield

$$\bar{\mu}_2=C\Phi(0)+\mu_0\,,\quad\text{and}\quad\bar{\mu}=(1-C)\bar{\mu}_2\,, \tag{3.30}$$

in which we interpret μ_0 as the moment on an isolated type 2 atom. Marshall (1968) includes other than random alloys and binary systems with moments on both constituents. In presenting the above example of the model we hoped simply to show the assumptions behind the model, the $\boldsymbol{\kappa}=0$ limits of the model and the variation of $\bar{\mu}$ versus C implicit in the model. This model is 'linear' in the sense that the perturbations at μ_j superpose but it is not linear in the sense of a straight line relationship between $\bar{\mu}$ and C.

At this point we will examine the wide set of results obtained on the Ni–Cu alloy system using both the purely magnetic and the polarization dependent cross sections. In some ways the Ni–Cu system is the prototype for ferromagnetic alloy systems and the principles tested out on it are probably more widely applicable. Aldred *et al.* (1973) investigated the moment distribution in Ni–Cu alloys across a wide band of ferromagnetic alloy concentrations. They measured the purely magnetic diffuse cross section and also the nuclear diffuse scattering from which they obtained the Cowley s.r.o. parameters for the alloys. The analysis of the magnetic scattering was first of all simplified by ignoring the higher order (i.e. s.r.o. fluctuation) terms in the Marshall cross section expression, although the effect of mean s.r.o. was retained in the dominant concentration fluctuation term. They further simplified the Marshall formula by neglecting any effect on the moment of copper atoms due to their environment whilst including a provision for copper sites to carry a moment. The final expression with which they fitted their data was then very similar to the first term in eqn (3.24) except provision for a nonzero mean short range order was made. It is instructive to look at the moment disturbance parameters $\phi(\boldsymbol{R}_i)$ obtained from fitting the modified Marshall formula to the data. In Fig. 3.1 we show the various $\phi(\boldsymbol{R}_i)$ for the first to fourth shells of neighbours. For most of the composition range the largest defect occurs on first neighbours and this defect is relatively independent of composition (i.e. crudely, the Marshall model works and the analysis is self consistent). There is, however, a definite breakdown of the Marshall model evident in the results. The second and further neighbour defects increase with concentration until they are comparable with the first neighbour defect at 40 at % copper. In the Marshall model $\Phi(\boldsymbol{\kappa})$, and thus the values of all disturbance parameters, is independent of composition. This is clearly not the case in the Ni–Cu alloy system. This variation of $\Phi(\boldsymbol{\kappa})$ with composition is confirmed in by other workers. In the work of Medina and Cable (1976) the polarization dependent cross section was measured, out of which $\Phi(\boldsymbol{\kappa})$ could be extracted unambiguously because of the simplicity of the expression for $\Delta(\mathrm{d}\sigma/\mathrm{d}\Omega)$. The data of Medina and Cable is also the most accurate available for the Ni–Cu system because they used samples containing the isotope ^{62}Ni which makes $(b_2 - b_1)$ in eqn (3.28) very large and $\Delta(\mathrm{d}\sigma/\mathrm{d}\Omega)$ relatively easy to measure. Use of the ^{62}Ni had a further useful function as $(b_2 - b_1)^2$ occurs in the nuclear defect cross section (see equation 3.6) which makes this cross section overwhelmingly larger than the purely magnetic cross section in an unpolarized neutron experiment. It is then easy to extract $\sum_\rho \alpha_\rho \exp(i\boldsymbol{\kappa}.\boldsymbol{\rho})$ from such an experiment. Medina and Cable used this to correct the data to that expected for a random alloy (Fig. 3.2). From this data it is clear that $\Phi(\boldsymbol{\kappa})$ is strongly dependent on concentration especially near the critical concentration for ferromagnetism. In this they agree with the earlier work of Hicks *et al.* (1969) which showed that near the critical concentration the static moment fluctuations have a very long range.

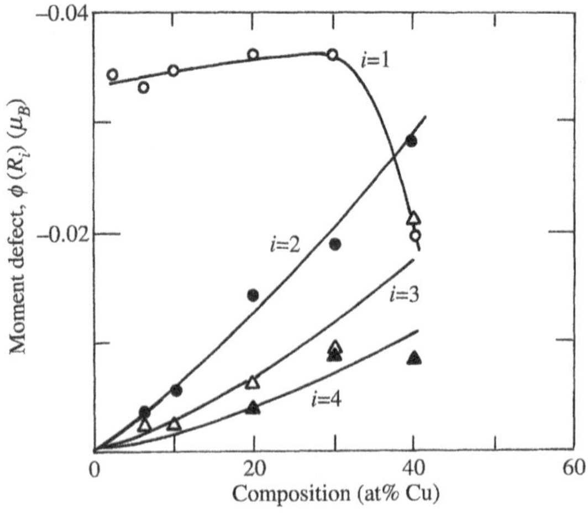

Fig. 3.1 The variation of first to fourth neighbour magnetic defect parameters with composition for Ni–Cu alloys. (After Aldred *et al.* 1983).

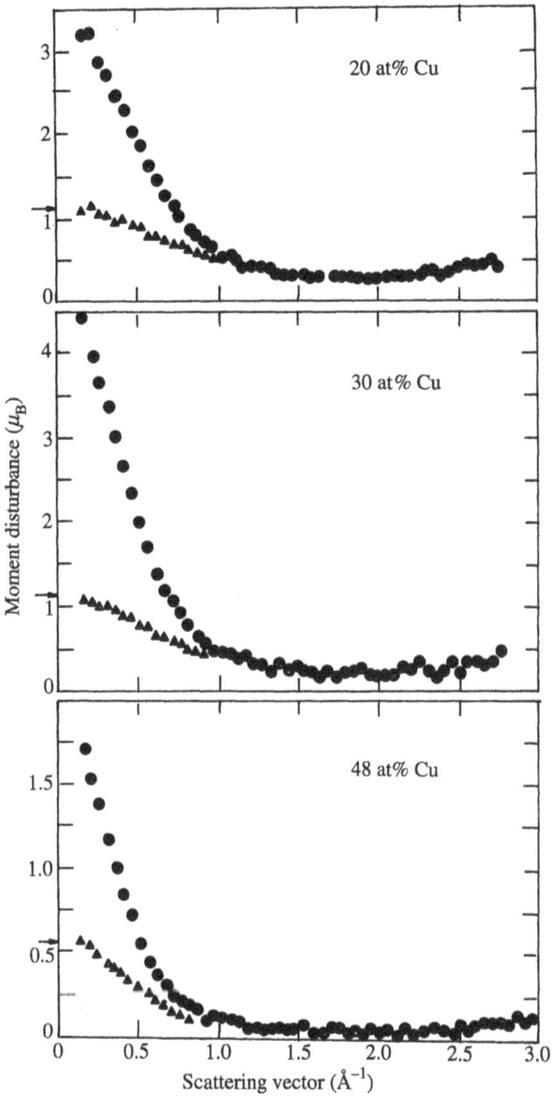

Fig. 3.2 The polarization dependent cross sections expressed in terms of moment disturbance for Ni–Cu alloys. (After Medina and Cable 1976.) The arrows indicate the measured $d\mu/dC$.

There is one other very simple reason why the Marshall model fails for the Ni–Cu system, and that is the occurrence of a critical concentration for ferromagnetism which is not a percolation concentration for the breakdown of coupling between individual atomic moments. Rather it appears to be crudely the composition at which the moment on the magnetic species disppears. The

Marshall model does not admit the possibility of such a concentration [compare eqn (3.30)]. Models which include the possibility of a complete loss of moment on the magnetic species (Medina and Cable 1977; Hicks 1977) have the common feature of including the effect of the *magnetic* environment as well as the *atomic* environment on the magnitude of each atomic moment in the system.

3.4 THE MAGNETIC ENVIRONMENT MODEL

The limitation of the Marshall model is the assumption of a magnetic defect independent of concentration. This limitation can be partially overcome (Balcar and Marshall 1968) by modifying the magnetic defect in the presence of a second defect centre. Thus if the disturbance produced on an atom of type 2 by a type 1 atom at a distance $\boldsymbol{R}$ is $\phi(\boldsymbol{R})$ the presence of type 1 atoms at both $\boldsymbol{R}$ and $\boldsymbol{T}$ will give rise to an extra influence on the magnetic moment which can be described by the function $\xi(\boldsymbol{R},\boldsymbol{T})$. In this way the inherent restriction in using only $\phi(\boldsymbol{R})$, that is that the change in magnetic moment of an atom is obtained by a *linear superposition* of the effects from neighbouring atoms, is removed. If the moment on the atom of type 2 is affected by further neighbours of type 1 a further function must be introduced for each extra neighbour, so that for long range defects or for alloys concentrated in type 1 atoms the expressions can become very cumbersome.

This is the case when an alloy is near its critical concentration for ferromagnetism. The static fluctuations of moment become very long ranged as can be seen in Fig. 3.1 for Ni–Cu. In this and similar cases a model which introduces a *mechanism* for the nonlinear moment disturbance has been widely used. This magnetic environment model assumes that the moment on an atom is a function of the moments on neighbouring atoms as well as their presence. In this way the truly cooperative nature of the approach to the critical concentration for ferromagnetism can be modelled. It should be emphasized that the distribution of magnetization in a ferromagnetic alloy is, in the end, simply a function of the distribution of the atoms in the alloy. The function of the magnetic environment model is to provide an analytical form for the effect of the atom distribution on the moment of each atom rather than relying on a parameter for the effect of the presence of each atom at its particular location. The magnetic environment model was first proposed by Hicks (1970) for Ni–Cu alloys near their critical concentration. The model was then developed by Medina and Cable (1977) and Hicks (1977).

The form developed by Medina and Cable (1977) approached the functional variation of atomic moment quite generally in the following way. The moment on an atom of type 1 is given by

$$\mu_1(\boldsymbol{R}) = F\{h(\boldsymbol{R}), \nu(\boldsymbol{R})\}$$

$$= \bar{\mu}_1 + \frac{\partial F}{\partial h}\{h(\boldsymbol{R}) - h_{eff}\} + \frac{\partial F}{\partial \nu}\{\nu(\boldsymbol{R}) - \langle\nu\rangle\}, \tag{3.31}$$

with

$$\overline{\mu}_1 = F(h_{eff}, \langle \nu \rangle),$$

$$h(\boldsymbol{R}) = \sum_{R' \neq 0} \mathscr{I}(\boldsymbol{R} - \boldsymbol{R}')\,\mu(R')$$

$$\nu(\boldsymbol{R}) = \sum_{\delta} p(\boldsymbol{R} + \delta)\,. \tag{3.32}$$

The last sum is over the δ first neighbours of $\boldsymbol{R}$ and $p = 1$ for an atom of type 1 at the first neighbour site and $p = 0$ for a type 2 atom. Equation (3.31) may then be substituted into either eqn (2.6) for unpolarized cross sections or into eqn (2.7) for polarized cross sections. Medina and Cable (1977) used a perturbation expansion of eqn (3.31) linear in the chemical and magnetic terms, and with no moment on type 2 atoms, to fit the observed cross sections and obtain the variation of the chemical and magnetic parameters with concentration for Ni–Cu alloys.

Hicks (1970) used a similar approach, but rather than assume a general functional dependence of moment on chemical and magnetic environments to be linearized later, assumed a simple function which saturates with exchange field and is linear in the occupation of surrounding sites. In this way Hicks (1977), could model together the cross sections and the variation of magnetic moment with concentration. In this model the magnetic moment on an atom is given by

$$\mu(\boldsymbol{R}) = \frac{\varsigma(\boldsymbol{R})p(\boldsymbol{R})h(\boldsymbol{R})}{1 + Bh(\boldsymbol{R})}\,, \tag{3.33}$$

where $p(\boldsymbol{R})$ is 0 or 1 for nonmagnetic and magnetic sites respectively so that there is only one magnetic species in the model, $\zeta(\boldsymbol{R})$ is the initial susceptibility of magnetic sites, and $\zeta(\boldsymbol{R})/\mathrm{B}$ is the high field moment which can be developed at $\boldsymbol{R}$. The variation with atomic environment is through the initial susceptibility

$$\zeta(\boldsymbol{R}) = a(0) + \sum_{R' \neq R} a(\boldsymbol{R} - \boldsymbol{R}')p(\boldsymbol{R}')\,, \tag{3.34}$$

so that with the addition of magnetic atoms at $\boldsymbol{R}'$, there are additions to the susceptibility $a(0)$ of a completely isolated magnetic atom at $\boldsymbol{R}$. The interaction between magnetic atoms is further represented by an exchange field generated by the moments at all other sites through an exchange interaction $\mathscr{I}(\boldsymbol{R} - \boldsymbol{R}')$ with

$$h(\boldsymbol{R}) = \sum_{R'' \neq R} \mathscr{I}(\boldsymbol{R} - \boldsymbol{R}'')\mu(\boldsymbol{R}'')\,. \tag{3.35}$$

Putting all the above together an integral equation is obtained.

$$M(\boldsymbol{\kappa}) = \frac{1}{NV_B}\int \mathrm{d}\boldsymbol{\kappa}' I(\boldsymbol{\kappa}')M(\boldsymbol{\kappa}')\left[\frac{1}{N\Omega_B}\int \mathrm{d}\boldsymbol{\kappa}''\left(a(0)\sum_{R}\exp(\mathrm{i}\boldsymbol{\kappa}''\cdot\boldsymbol{R}) + \right.\right.$$

$$\left.\left. A(\boldsymbol{\kappa}'')P(\boldsymbol{\kappa}'')\right) \times P(\boldsymbol{\kappa} - \boldsymbol{\kappa}' - \boldsymbol{\kappa}'') - BM(\boldsymbol{\kappa} - \boldsymbol{\kappa}')\right]\,, \tag{3.36}$$

where N is the number of sites in the sample and V_B is the volume of a Brillouin zone with all integrals over the Brillouin zone. $M(\boldsymbol{\kappa})$, $P(\boldsymbol{\kappa})$, $A(\boldsymbol{\kappa})$, and $I(\boldsymbol{\kappa})$ are the Fourier transforms of $\boldsymbol{\mu}(\boldsymbol{R})$, $p(\boldsymbol{R})$, $a(\boldsymbol{R})$, and $\mathscr{I}(\boldsymbol{R})$. This integral equation can be solved by breaking $M(\boldsymbol{\kappa})$ and $P(\boldsymbol{\kappa})$ into their average and fluctuating parts thus

$$M(\boldsymbol{\kappa}) = \bar{\mu}\sum_R \exp(\mathrm{i}\boldsymbol{\kappa}\cdot\boldsymbol{R}) + F(\boldsymbol{\kappa}) \quad \text{and}$$

$$P(\boldsymbol{\kappa}) = C\sum_R \exp(\mathrm{i}\boldsymbol{\kappa}\cdot\boldsymbol{R}) + D(\boldsymbol{\kappa}). \tag{3.37}$$

If certain convoluted terms are omitted the pair of solutions is

$$\bar{\mu} = \frac{I(0)C[a(0)+CA(0)]-1}{BI(0)} \qquad \text{and}$$

$$F(\boldsymbol{\kappa}) = \frac{\{\mathrm{d}\bar{\mu}/\mathrm{d}C - [A(0)-A(\boldsymbol{\kappa})]C/B\}D(\boldsymbol{\kappa})}{1+[I(0)-I(\boldsymbol{\kappa})]/I^2(0)B\bar{\mu}} \tag{3.38}$$

for the average moment and for the Fourier transform of the moment fluctuations from the mean. $F(\boldsymbol{\kappa})$ may be written more simply as

$$F(\boldsymbol{\kappa}) = \frac{\{\mathrm{d}\bar{\mu}/\mathrm{d}C - [A(0)-A(\boldsymbol{\kappa})]C/B\}D(\boldsymbol{\kappa})}{1+[I(0)-I(\boldsymbol{\kappa})]/I^2(0)B\bar{\mu}}. \tag{3.39}$$

The diffuse cross sections may be expressed in terms of the purely magnetic cross section

$$\frac{\mathrm{d}\sigma}{\mathrm{d}\Omega} = \left(\frac{e^2\gamma}{2m_e c^2}\right)^2 F(\boldsymbol{\kappa})F(-\boldsymbol{\kappa})$$

$$= \left(\frac{e^2\gamma}{2m_e c^2}\right)^2 C(1-C)S(\boldsymbol{\kappa})m(\boldsymbol{\kappa})m(-\boldsymbol{\kappa}), \tag{3.40}$$

with $m(\boldsymbol{\kappa})$ defined by $F(\boldsymbol{\kappa}) = m(\boldsymbol{\kappa})D(\boldsymbol{\kappa})$ and $D(\boldsymbol{\kappa})D(-\boldsymbol{\kappa}) = C(1-C)S(\boldsymbol{\kappa})$. $S(\boldsymbol{\kappa})$ is the Fourier transform of the Cowley atomic short range order parameters. Alternatively the polarized neutron difference cross section,

$$\Delta\frac{\mathrm{d}\sigma}{\mathrm{d}\Omega} = \left(\frac{e^2\gamma}{2m_e c^2}\right)(b_1-b_2)F(\boldsymbol{\kappa})D(-\boldsymbol{\kappa})$$

$$= \left(\frac{e^2\gamma}{2m_e c^2}\right)C(1-C)(b_1-b_2)S(\boldsymbol{\kappa})m(\boldsymbol{\kappa}) \tag{3.41}$$

may be written, with b_l and b_2 the scattering lengths of the two nuclear species. The cross sections at $\boldsymbol{\kappa} = 0$ are

$$\frac{d\sigma}{d\Omega} = \left(\frac{e^2\gamma}{2m_e c^2}\right)^2 C(1-C)S(0)\left(\frac{d\bar{\mu}}{dC}\right)^2$$

$$\Delta\frac{d\sigma}{d\Omega} = \left(\frac{e^2\gamma}{2m_e c^2}\right)(b_1 - b_2)C(1-C)S(0)\frac{d\bar{\mu}}{dC}, \tag{3.42}$$

where it is clear that the model returns the appropriate small $\boldsymbol{\kappa}$ limits apart from the term in the purely magnetic diffuse cross section which reflects the fluctuation of magnetization with the fluctuation of atomic short range order. This is a result of the neglect of certain convoluted terms in the solution for $\bar{\mu}$ and $F(\boldsymbol{\kappa})$.

3.5 APPLICATION OF THE MAGNETIC ENVIRONMENT MODEL

The magnetic environment model has been mostly applied to nickel alloys. Of the ferromagnetic 3d metals nickel is most faithfully described by an electron band model with no characteristics of atomic spin degrees of freedom. At low temperatures the moment at any site is then given by the excess of up over down electron spins which is a function of the Coulomb and exchange potentials.

The Ni–Cu system provides a simple and satisfactory example of the success of the model. The crucial composition region is that close to the ferromagnetic critical concentration where the magnetic cross sections peak in the forward scattering direction. For a simple comparison we can take a small $\boldsymbol{\kappa}$ form of $m(\boldsymbol{\kappa})$

$$m(\boldsymbol{\kappa}) = \frac{\left[d\bar{\mu}/dC - A(0)\,C\beta\boldsymbol{\kappa}^2/B\right]}{1 + \gamma\boldsymbol{\kappa}^2/I(0)B\bar{\mu}}, \tag{3.43}$$

in which the approximations

$$A(\boldsymbol{\kappa}) \approx A(0)\left[1 - \beta\boldsymbol{\kappa}^2\right] \quad \text{and}$$

$$I(\boldsymbol{\kappa}) \approx I(0)\left[1 - \gamma\boldsymbol{\kappa}^2\right] \tag{3.44}$$

are made. Because $A(\boldsymbol{\kappa})$ and $I(\boldsymbol{\kappa})$ are expanded to only second order $m(\boldsymbol{\kappa})$ depends only on the magnitude of $\boldsymbol{\kappa}$ in this approximation, and polycrystalline results in which $\boldsymbol{\kappa}$ is averaged over all directions can be used for a comparison with the expression (3.43). The constants $A(0)$, $I(0)$, and B can be obtained from a fit of the variation of $\bar{\mu}$ with C. The fit of eqn (3.38) to the data for Ni–Cu is shown in Fig. 3.3. Then if $A(\boldsymbol{\kappa})$, and $I(\boldsymbol{\kappa})$ are dominated by first neighbour interactions $\beta = \gamma = 1.05$ Å^{-1}, and $m(\boldsymbol{\kappa})$ is completely determined. Figure 3.4 shows $m^{-1}(\boldsymbol{\kappa})$ versus $\boldsymbol{\kappa}^2$ determined in this way with data obtained from polarized and unpolarized experiments. It is clear that the magnetic environment model contains the essential physics in this case. Similar results are obtained for Ni–Cr alloys and a similar success is obtained by application of the magnetic environment model (Hicks 1977).

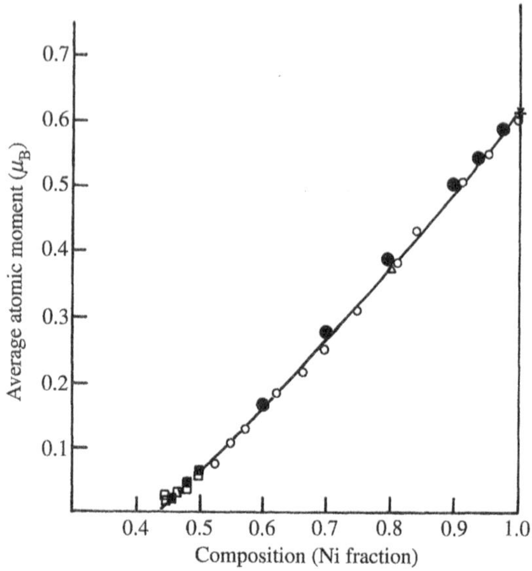

Fig. 3.3 The variation of average moment with nickel concentration for Ni–Cu. The full line is a fit of the magnetic environment model. The different symbols represent different experiments. (After Hicks 1977.)

The magnetic environment model can also be extended to systems in which both constituents carry a moment (Hicks 1980). The expression for the moment at any site $\boldsymbol{R}$ is an extension of eqn (3.32) using the same arbitrary but simple saturating function

$$\mu(\boldsymbol{R}) = \frac{\left[\zeta_2(\boldsymbol{R}) + p(\boldsymbol{R})\left(\zeta_1(\boldsymbol{R}) - \zeta_2(\boldsymbol{R})\right)\right]h(\boldsymbol{R})}{1 + \left[B_2 + p(\boldsymbol{R})(B_1 - B_2)\right]h(\boldsymbol{R})}, \tag{3.45}$$

where the subscripts refer to the two species and the symbols have their previous meanings. The initial susceptibilities are again simple superposing functions of their atomic environments and the exchange field $h(\boldsymbol{R})$ is given by

$$\begin{aligned} h(\boldsymbol{R}) = \sum_{\boldsymbol{R}''' \neq \boldsymbol{R}} \{ & p(\boldsymbol{R})p(\boldsymbol{R}''')\mathcal{J}_1(\boldsymbol{R} - \boldsymbol{R}''') \\ & + \left[p(\boldsymbol{R})\left(1 - P(\boldsymbol{R}''')\right) + \left(1 - p(\boldsymbol{R})\right)p(\boldsymbol{R}''')\right]\mathcal{J}_{12}(\boldsymbol{R} - \boldsymbol{R}''') \\ & + \left(1 - p(\boldsymbol{R})\right)\left(1 - p(\boldsymbol{R}''')\right)\mathcal{J}_2(\boldsymbol{R} - \boldsymbol{R}''')\}\mu(\boldsymbol{R}'''), \end{aligned} \tag{3.46}$$

where the subscripts delineate exchange interactions within and between the two species. The solution follows that for the one moment magnetic environment model and yields expressions for the average moment and the spatially fluctuating part of the moment but again ignores some convoluted terms in the

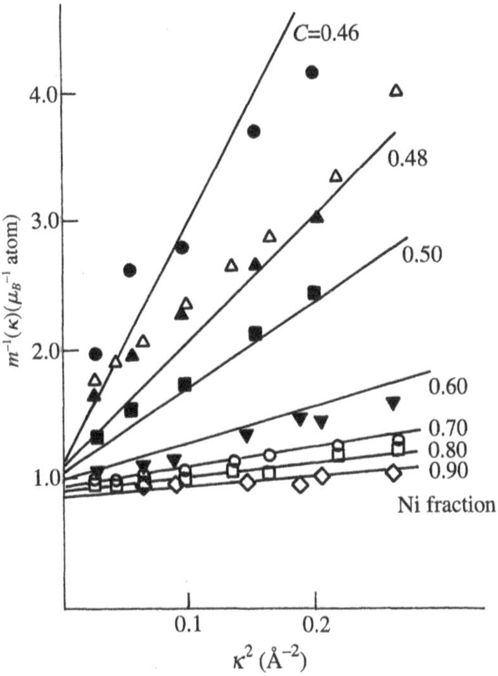

Fig. 3.4 The variation of inverse moment fluctuations with κ^2 for Ni–Cu alloys. Closed points are from unpolarized experiments and open points are from polarized experiments. The lines are parameter free predictions of the magnetic environment model assuming only first neighbour interactions. (After Hicks 1977.)

Fourier transforms which limit the moment fluctuations to be driven by composition fluctuations alone.

The model has been applied with success to the Ni–Pd, Ni–Rh (Hicks 1980), and Ni–Mn (Hicks and Moze 1981) alloy systems. The expression for the variation of average moment with composition now has seven parameters to be fitted; the three exchange interactions, the two parameters *B* which control the saturation functions, and three independent parameters which are combinations of the susceptibilities of the species and their variation with composition (Hicks and Moze 1981). This leaves one of the last undetermined from a fit of the variation of average moment with composition. The extra information can be obtained by using the data for the difference in species moments for just one composition. This can normally be estimated from the large $\boldsymbol{\kappa}$ cross section for polycrystalline alloys and compared with the $\boldsymbol{\kappa}$ independent part of the expression for the cross section from the model. Apart from the assumption of first neighbour only interactions there is no further information necessary to predict the full composition and $\boldsymbol{\kappa}$ dependence of the diffuse cross sections.

The Ni–Pd alloy system is interesting in that Cable and Child (1970) produced the unexpected result that addition of palladium to nickel increased the average moment on nickel. The magnetic environment model (Hicks 1980) supports this. The analysis in Hicks (1980) for the Ni–Pd system made some simplifications. Firstly, to reduce the number of parameters to be fitted it was assumed that $B_l = B_2$ because of the similarity of the two species. Secondly, the initial susceptibility for palladium isolated in nickel was arbitrarily put to zero. This latter can be remedied in this application of the magnetic environment model to Ni–Pd with a consequent even more striking agreement with the results of Cable and Child (1970). Rather than choose the isolated palladium susceptibility to be zero the redundancy can be taken out of the parameters by using the difference in moments between nickel and palladium as determined by Cable and Child (1970) for a 50 at% alloy. The curve (Fig. 3.5) fitted to the results for average moment remains exactly the same as in Hicks (1980).

Diffuse cross section data for Ni–Pd alloys come from unpolarized data of Aldred *et al.* (1970) as well as the polarized data of Cable and Child, 1970. The unpolarized data is concentrated at the palladium rich end of the composition range near the critical concentration for ferromagnetism and exhibits large low κ peak. In the polarized data, for much higher nickel concentrations, the peak in the diffuse cross section is away from the forward scattering direction. To demonstate the agreement between the magnetic environment model and both

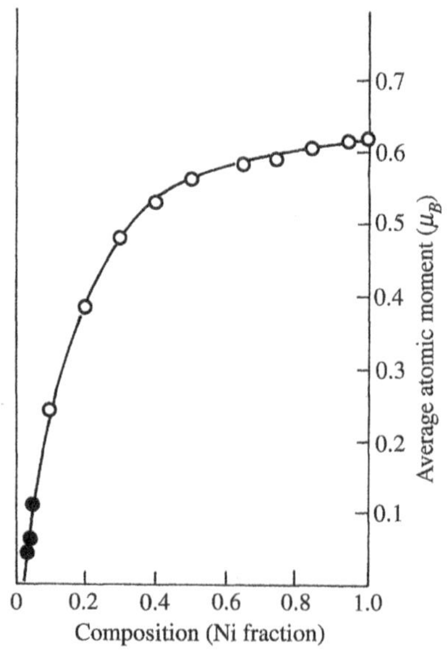

Fig. 3.5 The variation of average ferromagnetic moment with composition for Ni–Pd alloys. The line is the fit of the magnetic environment model to the data. (After Hicks 1980.)

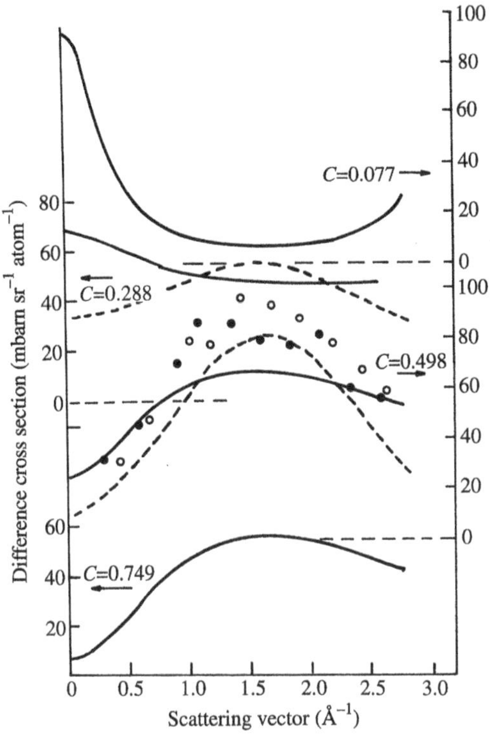

Fig. 3.6 The predicted polarized neutron difference cross sections for four polycrystalline Ni–Pd alloys under the assumption of first neighbour interactions only. The full curves are for random alloys. The broken curves are for alloys with first neighbour short range order and $S(0) = 0.5$. The experimental points are those of Cable and Child (1970). (After Hicks 1980.)

sets of data, it is necessary to calculate the predicted cross sections for more than the low $\boldsymbol{\kappa}$ region, and so orientation averages are needed for the higher $\boldsymbol{\kappa}$ region. Hicks (1980) used an average of over 63 independent single crystal orientations to simulate the polycrystalline results. Figure 3.6 shows the predictions for the polarized cross sections at the compositions examined by Cable and Child (1970). Unfortunately, no independent data exists for $S(\boldsymbol{\kappa})$ so this must be inferred from the polarized neutron data. Comparison of the $\boldsymbol{\kappa} = 0$ cross section with $\mathrm{d}\bar{\mu}/\mathrm{d}C$ implies that $S(0) \approx 0.5$ for the $C = 0.498$ alloy and the predictions for this alloy assume a first neighbour Cowley short range order parameter of −0.042. The same $S(\boldsymbol{\kappa})$ is used for the $C = 0.288$ alloy which produces a shallow maximum in the cross section similar to that observed by Cable and Child (1970) (not shown), but as the data for the $C = 0.749$ alloy suggests that that alloy is clustered rather than anticlustered no comparison is made because of the uncertainty in $S(\boldsymbol{\kappa})$.

For the palladium rich alloys we turn to the unpolarized results of Aldred *et al.* (1970). Again the results suggest that the alloys are anticlustered although because of the lack of data at very small scattering vectors no comparison with $d\bar{\mu}/dC$ is possible. Instead the curves calculated from the magnetic environment model and shown in Fig. 3.7 include the same short range order as that used for the more concentrated alloys. It can be seen that the agreement is good for all the comparisons, particularly as the data progresses from peaking in the forward scattering direction to peaking away from $\boldsymbol{\kappa} = 0$.

The magnetic environment model provides a prediction of all the spatial Fourier components of the moment distribution from the variation of average moment with composition, and this compares well with experiment. Of particular interest, however, is the average moment on each species. These can be obtained from combining the average moment with the difference in average moments of the two species. Further, the difference in the average moments is the $\boldsymbol{\kappa}$ independent part of $M(\boldsymbol{\kappa})$ which can be determined from the parameters used to fit the average moment and one other usually taken from the difference in average moments at a particular composition. From this information the average moments of the two species may be calculated for all compositions. The

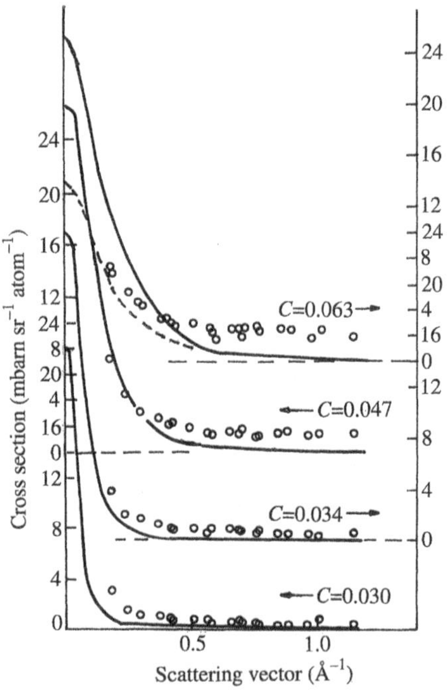

Fig. 3.7 The predicted unpolarized cross sections for polycrystalline Ni–Pd alloys close to the critical concentration. The full curves are for random alloys and the broken curve is for first neighbour short range order and S(0)=0.5. (After Hicks 1980.)

data on species moments from Hicks (1980) is shown in Fig. 3.8 along with curves which predict the moments based on the variation of average moment with composition and the difference in species moments at $C = 0.5$.

The magnetic environment model in its simple form does not take into account any relationship between the average moment and the degree of atomic short range order. Terms containing this effect were eliminated from eqn (3.35) to facilitate its solution. Evidence for effects like this is in the application of the magnetic environment by Hicks and Moze (1981) to the ferromagnetic Ni–Mn system. Above 20 at% manganese there is much evidence that the spontaneous moment of this system is extremely dependent on atomic short range order. The predictions of Hicks and Moze (1981) extrapolated as they must to the slope of the fitted variation of average moment with composition, however the data of Cable and Child (1974) did not. A possible explanation is that as the atomic short range order in the alloys is probably changing as the composition approaches that of Ni_3Mn, the observed change in average moment is entirely due to the change in composition at constant short range order. In circumstances like this the $\boldsymbol{\kappa} = 0$ value of the difference in the polarized neutron cross sections could well be different from the observed moment variation with composition, contaminated as it may be by an accompanying change in atomic short range order.

So the magnetic environment model for predicting the magnetic diffuse scattering works well for nickel alloys. It is expected to work less well for cobalt and iron alloys because of possible complications with the emerging vector nature of

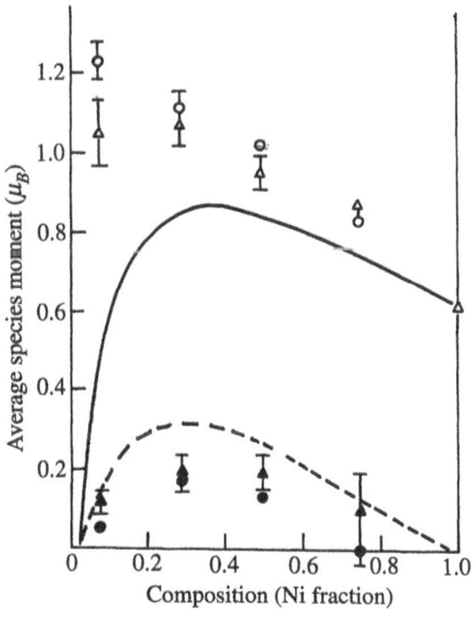

Fig. 3.8 Average moments of the individual species in Ni–Pd. The full curve is the nickel moment and the broken curve is the palladium moment predicted from the magnetic environment model. (After Hicks 1980.)

the spins involved although currently there are no studies reported. Apart from terms describing the moment fluctuations with fluctuations in atomic short range order the magnetic environment reduces to the Marshall model in the limit of infinite exchange field. In this way it has a similar relationship to the Marshall model as the Clapp and Moss (1966) model of atomic short range order has to the Cowley parameter description of atomic short range order, i.e. the replacement of correlation parameters with parameters with some physical significance.

3.6 THE AVERAGE MOMENT DISTRIBUTION IN FERROMAGNETIC ALLOYS

We have left this subject till last because it is the aspect of the distribution of magnetic moment in disordered alloys which can be directly compared with calculations of the electron band structure. This is possible because a simple version of the coherent potential approximation (CPA), that which includes only two different potentials for the electrons corresponding to the two atomic species, is sufficient to describe the average moment distribution. For a more complete description a distribution of potentials on each species corresponding to the distribution of magnetic moment would be needed rather than the substitution of the average environment.

The simple two potential version of the CPA can be expressed using a Hamiltonian of the form

$$\mathcal{H} = W + \sum_{nl\sigma_e} |nl\sigma_e > \epsilon_{nl\sigma_e} < nl\sigma_e|, \tag{3.47}$$

where the localized orbitals are given by

$$|nl\sigma_e > = N^{-1/2} \sum_{k_e} |nk_e\sigma_e > \mathrm{e}^{\mathrm{i}\kappa\cdot R_l}. \tag{3.48}$$

Here $\epsilon nl\sigma$ are the atomic energy levels which depend on the atom occupying site l, and $|nk_e\sigma_e\rangle$ are the electron Bloch orbitals. To fulfil the CPA condition that configurationally averaged electron scattering T-matrix should be zero

$$Ct_{1n\sigma_e} + (1-C)t_{2n\sigma_e} = 0 \tag{3.49}$$

(Soven 1967), where the ts are the single atom T-matrices given by

$$t_{1n\sigma_e}(z) = \left[\epsilon_{1n\sigma_e} - \Sigma_{n\sigma_e}\right]\left\{1 - F_{n\sigma_e}(z)\left[\epsilon_{1n\sigma_e} - \Sigma_{n\sigma_e}(z)\right]\right\}^{-1}, \tag{3.50}$$

with

$$F_{n\sigma_e} = -\frac{1}{N}\sum_k \langle nk\sigma_e|G(z)|nk\sigma_e\rangle = \int_{-\infty}^{\infty} \frac{\xi^{(0)}_{n\sigma_e}(\epsilon)}{z - \epsilon - \Sigma_{n\sigma_e}(z)}\mathrm{d}\epsilon \tag{3.51}$$

where $\xi^{(0)}_{n\sigma}(\epsilon)$ is the density of states generated by the periodic part of the Hamiltonian and z is the complex integration variable. The self energy can be obtained self consistently by means of eqns (3.49)–(3.51) written in the form

$$\Sigma(z) = \sum_{nl\sigma_e} | nl\sigma_e > \Sigma_{n\sigma_e}(z) < nl\sigma_e | . \tag{3.52}$$

The input quantities for this calculation are W, or a model for the density of states, and $\epsilon_{nl\sigma}$ the atomic energy levels for the two atom species. As well as the self energy the calculation gives the occupation numbers of the bands included in the model. In the case of transition metal alloys, eight bands incorporating the orbitals E_g and T_{2g} have been used (Leoni and Sacchetti 1974). This then can yield the average spin component of the ferromagnetic moment and the average distribution of the spin moment around atomic sites. Figure 3.9 shows the results of the application of the simple CPA to the Ni–Cu alloy system (Sacchetti, 1976). Although the average moment is reproduced quite well, including the critical concentration for ferromagnetism, the predicted average moment on Cu is significantly nonzero. The simple CPA has more success in predicting the symmetry of the average atomic moment distribution.

The symmetry of the moment distribution consists of a spherical part and a part described by a fourth order spherical Bessel function which is appropriate for the cubic symmetry. The simplest characterization of the spin moment distribution is the ratio of the aspherical part to the spherical part and this, as it is related to the ratio of the populations of the E_g and T_{2g} orbitals, can be used as a test of the simple CPA. The most sensitive experiments are those in which $\bar{\mu}$ and $f(\boldsymbol{\kappa})$ are determined from the first term in eqn (2.7) which leads to the difference in Bragg cross sections

$$\Delta\frac{\mathrm{d}\sigma}{\mathrm{d}\Omega} = 4\left(\frac{e^2\gamma}{2m_e c^2}\right) f(\boldsymbol{\kappa}) \sum_{i,j} \bar{b}\bar{\mu}_\perp \mathrm{e}^{\mathrm{i}\boldsymbol{\kappa}\cdot(R_i - R_j)} . \tag{3.53}$$

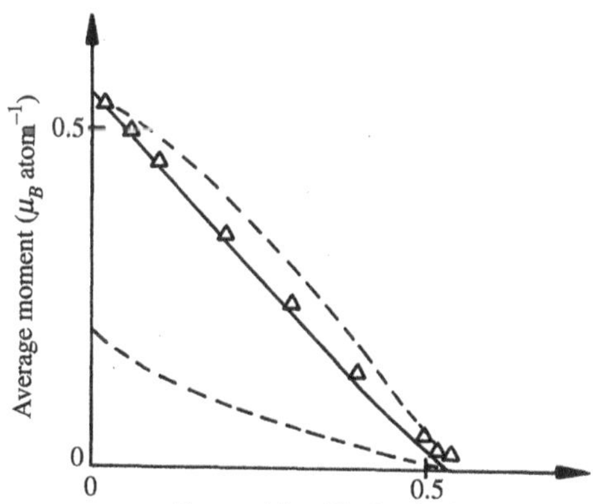

Fig. 3.9 Spin magnetic moments as a function of composition in Ni–Cu from simple CPA. The full line is the average moment and the broken lines the moments on nickel and copper. (After Sacchetti 1976.)

between incident polarization up and down. The increase in accuracy comes from the factor four, the generally larger value for nuclear compared with magnetic scattering lengths, and the appearance of $f(\boldsymbol{\kappa})$ rather than $f^2(\boldsymbol{\kappa})$ in the cross section. Figure 3.10 shows the amplitude of the aspherical part of the form factor taken from single crystal polarized neutron experiments compared with the predictions of the simple CPA. The CPA does much better than the rigid band model. Many absolute measurements of the form factor exist for a large range of ferromagnetic alloys however CPA calculations of the extent of the spherical and aspherical parts of the form factors for random binary alloys are sparse and most of the comparisons have been with the ratio of aspherical to spherical parts.

The emphasis of this chapter has been largely phenomemological, necessarily so because of the lack of first principle calculations of the quantities measured. Models like that of Marshall, and the magnetic environment model, will no doubt play a major part in the description of the moment distribution in ferromagnetic alloys for a long time to come because of the major effort of doing the band calculations.

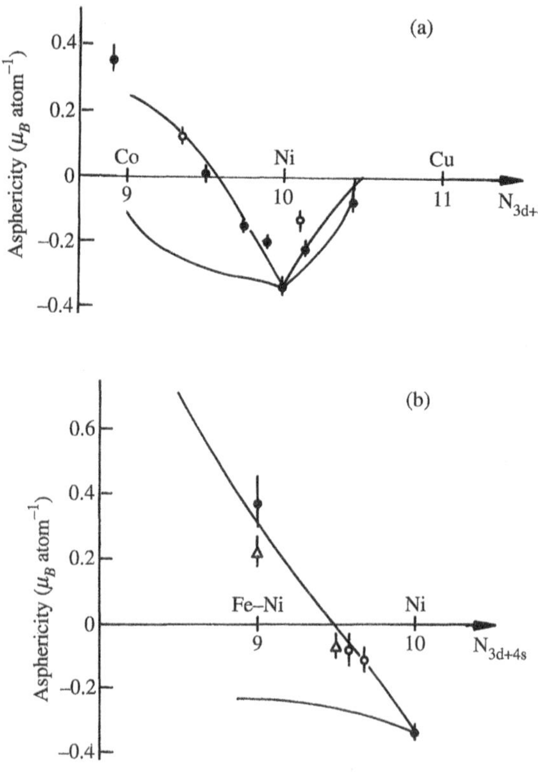

Fig. 3.10 Asphericity for various alloys. Comparison of experiments with the CPA for Co–Ni and Cu–Ni alloys (a) and Fe–Ni (b). Also shown is the curve for the rigid band model. (After Menzinger and Sacchetti 1976.)

References

Aldred, A.T., Rainford, B.D., and Stringfellow, M.W. (1970). *Phys. Rev. Lett.* **24**, 897.

Aldred, A.T., Rainford, B.D., Hicks, T.J., and Kouvel, J.S. (1973). *Phys. Rev. B* **7**, 218.

Balcar, E. and Marshall, W. (1968). *J. Phys. C* **1**, 966.

Cable, J.W. and Child, H.R. (1970). *Phys. Rev. B* **1**, 3809.

Cable, J.W. and Child, H.R. (1974). *Phys. Rev. B* **10**, 4607.

Clapp, P. and Moss, S.C. (1966). *Phys. Rev.* **142**, 418.

Hicks, T.J., Rainford, B.D., Kouvel, J.S., Low, G.G., and Comly, J.B. (1969). *Phys. Rev. Lett.* **22**, 531.

Hicks, T.J. (1970). *Phys. Lett.* **32A**, 410.

Hicks, T.J. (1977). *J. Phys. F* **7**, 481.

Hicks, T.J. (1980). *J. Phys. F* **10**, 879.

Hicks, T.J. and Moze O. (1981). *J. Phys. F* **11**, 2633.

Leoni, F. and Sacchetti, F. (1974). *Nuovo Cim.* **21B**, 97.

Marshall, W. (1968). *J. Phys. C* **1**, 88.

Medina, R.A. and Cable, J.W. (1976). *AIP Conf. Proc.* **290**, 292.

Medina, R.A. and Cable, J.W. (1977). *Phys. Rev. B* **15**, 1539.

Medina, R.A. and Garland, J.W. (1976). *Phys. Rev. B* **14**, 5060.

Menzinger, F. and Sacchetti, F. (1976). *Proceedings of the conference on neutron scattering*, Gatlinburg, TN, June 1976, p. 677.

Sacchetti, F. (1976). *Nuovo Cim.* **32B** 285.

Shull, C.G. and Wilkinson, M.K. (1955). *Phys. Rev.* **97**, 304.

Soven, P. (1967). *Phys. Rev.* **156**, 809.

Staunton, J.B., Johnson, D.D., and Pinski, F.J. (1990). *Phys. Rev. Lett.* **65**, 1259.

Staunton, J.B., Gyorffy, B.L., Johnson, D.D., Pinski, F.J. and Stocks G.M. (1989). Local density theory of magnetism and its interrelation with compositional order in alloys. In *Alloy phase stability*, (eds. G.M. Stocks and A. Gonis) p. 469. Kluwer Academic Publishers.

4

Defect scattering from antiferromagnetic alloys

Antiferromagnetic alloys of the 3d transition series are interesting for the same reason as their ferromagnetic cousins. The problems of measurement of the magnetic defect cross section is, however, quite different. Whereas for the ferromagnetic systems it was possible to separate out the magnetic defect cross section either by manipulating the ferromagnetic direction with a field or by measuring total cross sections for incident polarized neutrons with the polarization along or opposite to the ferromagnetic direction, antiferromagnetic alloys in general do not have such a direction.

To definitely isolate the magnetic defect cross section then involves analysis of the polarization of the scattered neutrons. The most popular means of doing this has been to align the neutron polarization along the scattering vector. From eqn (1.15) for a collinear antiferromagnet $\boldsymbol{M}(\boldsymbol{\kappa})$ has an axis but no direction so that the only term which can change the polarization is the first of the purely magnetic terms

$$-\left(\frac{e\gamma}{\hbar c}\right)^2 \overline{\langle \boldsymbol{M}_\perp^*(\boldsymbol{\kappa})\rangle \cdot \langle \boldsymbol{M}_\perp(\boldsymbol{\kappa})\rangle \boldsymbol{P}}\,, \tag{4.1}$$

which as can be seen reverses the polarization and provides a definitive separation of the purely magnetic scattering. Unfortunately, it is not the only term which provides some reversal of the polarization; nuclear spin incoherent scattering also does, as can be seen from eqn (1.20). However the nuclear spin scattering is isotropic and is generally well known, so that its effect can be subtracted. Manipulation of the incident polarization direction can eliminate the nuclear spin incoherent scattering as that part of the nuclear spin cross section which reverses the polarization is independent of the polarization direction. On the other hand, change in incident polarization direction *always* alters the fraction of purely magnetic scattering which occurs with polarization change.

Triple axis spectrometers can always be adapted to polarization analysis measurements by insertion of polarizing crystals at the monochromating and analysing axes and many measurements of diffuse magnetic scattering have been done using triple axis spectrometers in this way. However in the next section we highlight two diffractometer-spectrometers which have been specifically designed with diffuse scattering measurements with polarization analysis in mind.

4.1 POLARIZATION ANALYSIS INSTRUMENTS FOR DIFFUSE SCATTERING

The immediate ancestor of the D7 instrument at the Institut Laue–Langevin was a diffuse scattering instrument on the cold source of the reactor at Jülich. The emphasis was on intensity because of the very low diffuse cross sections and the instrument used a mechanical velocity selector to provide a wavelength which could be varied from 4–20 Å with a FWHM of 15 per cent. The detectors had only a light shielding of boron carbide but surrounding the whole instrument there was a wooden cylinder lined with cadmium on the inside. Sixty-three detectors were located on mobile banks to cover a wide band of scattering angles. Sometimes a Fermi chopper was located before the specimen and time of flight spectra were obtained to isolate the elastic scattering or to locate inelastic components of the scattering.

The same philosophy of diffuse scattering instrument design was carried over into D7 which has relaxed resolution, and multidetector banks mounted inside a cylindrical enclosure which is the main neutron shield. Figure 4.1 shows the layout of D7. Instead of a mechanical monochromator, D7 has a triple pyrolitic

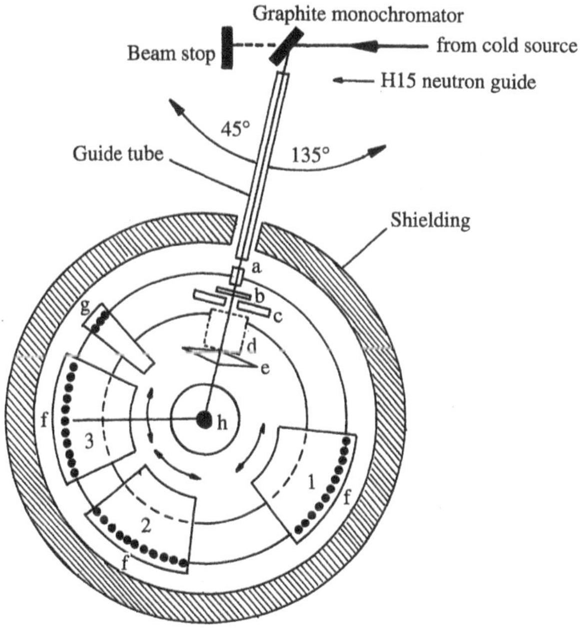

Fig. 4.1 The layout of D7 at the Institut Laue–Langevin. (a) Be filter or polarizer, (b) main monitor, (c) slit system, (d) flipper and second polarizer, (e) disk chopper or flipper, (f) detectors with supermirror analysers, (g) vertical detector bank, (h) sample. (After Maier 1983.)

graphite monochromator inserted in the guide tube at postion H15 in the original guide hall at the Institut Laue–Langevin. The three elements in the monochromator allow three different wavelengths to be selected without altering anything in the guide tube, thus avoiding an interruption to experiments downstream on the guide. The relaxed wavelength resolution of 3 per cent is matched by the angular width of the detectors. What distinguishes D7 from its predecessors is its ability to separate magnetic diffuse scattering from nuclear diffuse scattering using polarization analysis. No other measurement involving neutron scattering needs this separation more than diffuse scattering. In both Bragg scattering and scattering involving sharp excitations there is normally circumstantial evidence to identify the magnetic features which are well resolved from all others. In diffuse scattering the magnetic usually overlays the nuclear scattering and a definite experimental separation is needed. For polarization and analysis D7 uses 'supermirrors' (Schärpf 1982). These supermirrors are deposited on thin glass bent into curved Soller guides of length 300 mm. Within the supermirrors the layers are vanadium and a Co–Fe alloy chosen so that the refractive index for one neutron spin direction is unity in both types of layer. Neutrons with this spin are then transmitted undeviated through the layers to be absorbed in a layer next to the glass, again with refractive index unity, but with high absorption (Schärpf 1982). A π-type spin flipping coil is used to flip the neutrons for the two conditions required to separate non spin flip and spin flip cross sections. Polarization is initially vertical but its direction at the specimen postion can be varied by three orthogonal Helmholtz coil pairs located there and large enough to accommodate cryostats and other devices. This enables the proportion of the magnetic scattering going into the spin flip cross section to be varied.

By locating a second polarizing supermirror after the flipper, time of flight spectroscopy can be done. With spin flipper off the beam is transmitted, but with spin flipper on the beam intensity is reduced to a low level which depends on the flipping ratio of the polarizers. This device can then be used as a simple Fermi chopper, or alternatively it can be used as statistical chopper by feeding a psuedorandom pulse sequence to the flipper. The latter has the advantage of a higher mark space ratio and therefore intensity, but the disadvantage of a 'background of ignorance'. Instead of the flipper chopper a conventional mechanical Fermi chopper can also be used.

LONGPOL is another diffuse scattering instrument incorporating polarization analysis which in its various forms has been located on the HIFAR reactor at Lucas Heights. LONGPOL has used polycrystalline iron in a saturating magnetic field for both polarization and analysis. The iron filters are a cheap technology which has the disadvantage of poor polarization (from 40 to 50 per cent for neutrons with a wavelength of 3.6 Å) and high fringing fields. It is, however, possible to optimize the combination of intensity and polarization to minimize the error by variation of the filter thickness. The iron filters also have the advantage of a wide angle of acceptance; very important especially on the analysis side for an instrument with relaxed angular resolution. LONGPOL's wavelength resolution is quite relaxed at a FWHM of 8 per cent and this has been achieved by a double monochromator arrangement with pyrolitic graphite with a mosaic spread of 3.5° operating at a wavelength of 3.6 Å. Figure 4.2 shows the configuration of LONGPOL used for many of the experiments on antiferromagnetic defect scattering and scattering from spin glasses. The neutron polarization direction was kept along the scattering vector so that magnetic scattering was with reversal of polarization. More recently LONGPOL has been rebuilt with eight detectors rather than one.

For much of its life LONGPOL has incorporated time of flight spectroscopy mainly in order to confirm that the scattering being measured is elastic or closely elastic. The neutron time of flight has been measured by pulsing the spin flipper with a psuedorandom sequence and cross correlating the resulting time sequence at the detector with that impressed on the spin flipper. The resulting time of flight

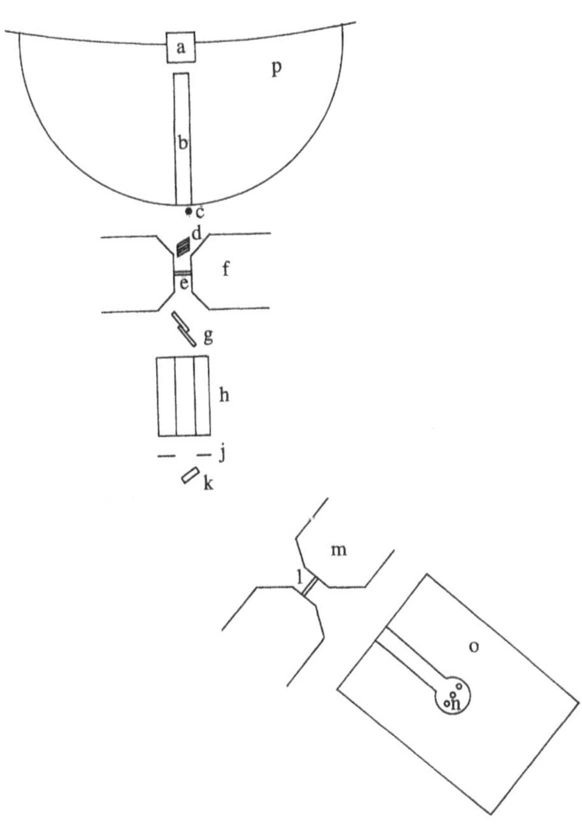

Fig. 4.2 Plan drawing of LONGPOL main components. (a) Monochromator, (b) collimator, (c) monitor, (d) λ/n filters, (e) polarizing filter, (f) magnet, (g)$\lambda/2$ filter, (h) spin flipper, (j) Cd aperture, (k) sample, (l) analysing filter, (m) magnet, (n) detector, (o) detector shielding, (p) pile shielding.

spectrum is very noisy and this can mask small inelastic features but the intensity associated with the elastic peak is normally well defined. Figure 4.3 shows a time of flight spectrum from a single crystal of antiferromagnetic Mn–Ni (Davis *et al.* 1982). The elastic peak is quite obvious but a small magnon peak is also seen. The spectrum was taken with the neutron polarization direction essentially along the scattering vector so that the magnon scattering was largely with neutron spin flip. The elastic peak however was predominantly without neutron spin flip as indicated by its relative sign.

4.2 MODELS FOR ANTIFERROMAGNETS

The extension of the Marshall model (Marshall 1968) to antiferromagnets is quite straightforward. Firstly we define the staggered moment

$$\nu_i = \mu_i \exp(\mathrm{i}\boldsymbol{\tau} \cdot \boldsymbol{R}_i), \text{ with}$$

$$\langle \mu_i \rangle = 0, \text{ and}$$

$$\langle \nu_i \rangle = \bar{\nu}, \tag{4.2}$$

in which $\boldsymbol{\tau}$ is the shortest reciprocal lattice vector of the antiferromagnetic structure and $\exp(\mathrm{i}\boldsymbol{\tau} \cdot \boldsymbol{R}_i)$ is +1 or −1 depending on whether $\boldsymbol{R}_i$ belongs to one sublattice or the other. This gives

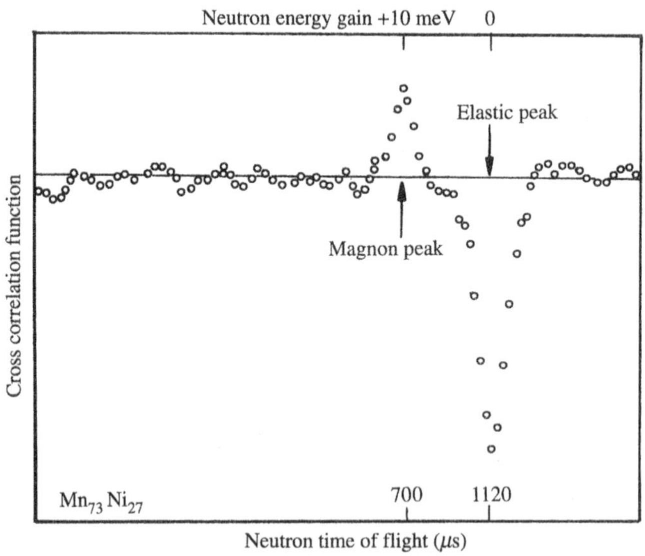

Fig. 4.3 Time of flight spectrum from antiferromagnetic γ-Mn–Ni. The elastic diffuse scattering peak and a magnon peak are visible. (After Davis *et al.* 1982.)

$$\frac{d\sigma}{d\Omega} = \left(\frac{e^2\gamma}{2m_e c^2}\right)^2 f^2(\boldsymbol{\kappa}) \sum_{ij} e^{i(\boldsymbol{\kappa}-\boldsymbol{\tau})\cdot(\boldsymbol{R}_i-\boldsymbol{R}_j)} \overline{(\nu_{i\perp} - \bar{\nu}_\perp)(\nu_{j\perp} - \bar{\nu}_\perp)} \tag{4.3}$$

for the purely magnetic scattering with the magnetic moment in Bohr magnetons.

Rather than write down the equations for one magnetic species as was done for ferromagnetic alloys (p. 32) the full dependence for two magnetic species is included here (Marshall 1968). It is assumed that the moment on each species is a linear function of its environment so that the staggered moment at any site is given by

$$\nu_i = \nu_2 + \sum_{\boldsymbol{\rho}} g(\boldsymbol{\rho})(p_{i+\boldsymbol{\rho}} - C) + p_i(\nu_1 - \nu_2) + \sum_{\boldsymbol{\rho}} p_i(p_{i+\boldsymbol{\rho}} - C)\{h(\boldsymbol{\rho}) - g(\boldsymbol{\rho})\} \tag{4.4}$$

in which $g(\boldsymbol{\rho})$ is the change from the average in the moment of an atom of type 2 at i due to the presence of an atom of type 1 at $i + \boldsymbol{\rho}$, and $h(\boldsymbol{\rho})$ is the equivalent parameter for an atom of type 1 in the presence of atoms of type 2. $p_i = 1$ if an atom of type 1 is at site i and it is zero if an atom of type 2 is at i. Substitution of eqn (4.4) into eqn (4.3) for a random alloy gives

$$\frac{d\sigma}{d\Omega} = \left(\frac{e^2\gamma}{2m_e c^2}\right)^2 NC(1-C)f^2(\boldsymbol{\kappa})\Big[\{\nu_{1\perp} - \nu_{2\perp} + (1-C)G_\perp(\boldsymbol{\kappa}-\boldsymbol{\tau}) + CH_\perp(\boldsymbol{\kappa}-\boldsymbol{\tau})\}^2 + C(1-C)\{Q_\perp(0) - Q_\perp(\boldsymbol{\kappa}-\boldsymbol{\tau})\}\Big], \tag{4.5}$$

with

$$Q_\perp(\boldsymbol{\kappa}-\boldsymbol{\tau}) = \sum_{\boldsymbol{\rho}} e^{i(\boldsymbol{\kappa}-\boldsymbol{\tau})\cdot\boldsymbol{\rho}}\{h_\perp(\boldsymbol{\rho}) - g_\perp(\boldsymbol{\rho})\}^2,$$

and with G and H the Fourier transforms of g and h. Although the mean short range order is assumed to be zero in this expression the term proportional to $C^2(1-C)^2$ is due to spatial fluctuations in short range order. This expression can also be compared at $\boldsymbol{\kappa} = \boldsymbol{\tau}$ with the variation of the bulk staggered moment with composition, which in contrast with the ferromagnetic case must now be determined from the magnetic Bragg peaks using neutron diffraction. Using eqn (4.3) above

$$\frac{d\nu}{dC} = \nu_1 - \nu_2 + (1-C)G(0) + CH(0), \tag{4.6}$$

which when compared to the first term of eqn (4.5) is the same as for the ferromagnetic case when the spatial fluctuations in moment are driven by composition fluctuations. The distribution of magnetic moment in an antiferromagnetic alloy can then be described in terms of $g(\boldsymbol{\rho})$ and $h(\boldsymbol{\rho})$.

The magnetic environment model can also be written down for the antiferromagnetic case (Davis and Hicks 1981). The moment at any site $\boldsymbol{R}$ is written

$$\nu(\boldsymbol{R}) = \frac{[\zeta_2(\boldsymbol{R}) + p(\boldsymbol{R})\{\zeta_1(\boldsymbol{R}) - \zeta_2(\boldsymbol{R})\}]h(\boldsymbol{R})}{1 + \exp(\mathrm{i}\boldsymbol{\tau}\cdot\boldsymbol{R})[B_2 + p(\boldsymbol{R})(B_1 - B_2)]h(\boldsymbol{R})}, \tag{4.7}$$

with $\exp(\mathrm{i}\boldsymbol{\tau}\cdot\boldsymbol{R}) = \pm 1$ depending on which sublattice the site $\boldsymbol{R}$ is on, as above, and $\boldsymbol{\tau}$ is the shortest reciprocal lattice vector of the antiferromagnetic structure. This retains the saturating properties of the function for both signs of $h(\boldsymbol{R})$ which occur for an antiferromagnet, that is $\exp(\mathrm{i}\boldsymbol{\tau}\cdot\boldsymbol{R})[B_2 + p(\boldsymbol{R})(B_1 - B_2)]h(\boldsymbol{R})$ will always remain positive. $\zeta_1(\boldsymbol{R})$ and $\zeta_2(\boldsymbol{R})$ are the intitial susceptibilities of the two atom types and $\zeta_1(\boldsymbol{R})/B_1$ and $\zeta_2(\boldsymbol{R})/B_2$ are their saturating moments as for the two moment ferromagnetic case. The binary antiferromagnetic alloy then consists of moments at sites which depend on the species with its atomic environment and which see an exchange field which depends on the atomic and magnetic environments. The initial susceptibilities are assumed to be linear functions of their environment;

$$\zeta_1(\boldsymbol{R}) = a_1(0) + \sum_{\boldsymbol{R}'\neq\boldsymbol{R}} a_1(\boldsymbol{R}-\boldsymbol{R}')p(\boldsymbol{R}')$$

$$\zeta_2(\boldsymbol{R}) = a_2(0) + \sum_{\boldsymbol{R}''\neq\boldsymbol{R}} a_2(\boldsymbol{R}-\boldsymbol{R}'')(1-p(\boldsymbol{R}'')), \tag{4.8}$$

and the exchange field at $\boldsymbol{R}$ may be written

$$\begin{aligned} h(\boldsymbol{R}) = \sum_{\boldsymbol{R}'''\neq\boldsymbol{R}} \{ & p(\boldsymbol{R})p(\boldsymbol{R}''')\mathcal{J}_1(\boldsymbol{R}-\boldsymbol{R}''') + \\ & [p(\boldsymbol{R})(1-p(\boldsymbol{R}''')) + (1-p(\boldsymbol{R}))p(\boldsymbol{R}''')]\mathcal{J}_{12}(\boldsymbol{R}-\boldsymbol{R}''') + \\ & (1-p(\boldsymbol{R}))(1-p(\boldsymbol{R}'''))\mathcal{J}_2(\boldsymbol{R}-\boldsymbol{R}''')\}\mu(\boldsymbol{R}'''). \end{aligned} \tag{4.9}$$

For a simple collinear antiferromagnet the moment at each site is

$$\nu(\boldsymbol{R}) = \mu(\boldsymbol{R})\exp(\mathrm{i}\boldsymbol{\tau}\cdot\boldsymbol{R}) \tag{4.10}$$

so that

$$\langle\mu(\boldsymbol{R})\rangle = 0 \quad \text{and} \quad \langle\nu(\boldsymbol{R})\rangle = \bar{\nu},$$

where $\bar{\nu}$ is the average alloy moment. The fluctuations of the staggered moment from the average, and fluctuations of the concentration from the average are written

$$F(\boldsymbol{\kappa}-\boldsymbol{\tau}) = \sum_{\boldsymbol{R}} (\nu(\boldsymbol{R}) - \bar{\nu})\mathrm{e}^{\mathrm{i}(\boldsymbol{\kappa}-\boldsymbol{\tau})\cdot\boldsymbol{R}}, \tag{4.11}$$

and

$$D(\boldsymbol{\kappa}-\boldsymbol{\tau})=\sum_{R}\left(p(\boldsymbol{R})-C\right)\mathrm{e}^{\mathrm{i}(\boldsymbol{\kappa}-\boldsymbol{\tau})\cdot\boldsymbol{R}}$$

Upon Fourier transforming eqn (4.7) two equations are obtained, one for the variation of average staggered moment with concentration

$$\overline{\nu}=\frac{I(\boldsymbol{\tau})[a(0)+A(0)]-1}{BI(\boldsymbol{\tau})},\tag{4.12}$$

and one for the spatial fluctuations of staggered moment

$$F(\boldsymbol{\kappa}-\boldsymbol{\tau})=\left\{\left[\frac{\mathrm{d}\overline{\nu}}{\mathrm{d}C}-\left(\frac{\mathrm{d}I(\boldsymbol{\tau})}{\mathrm{d}C}-\frac{\mathrm{d}I(\boldsymbol{\kappa})}{\mathrm{d}C}\right)\frac{1}{2BI^2(\boldsymbol{\tau})}-\right.\right.$$
$$\left.\left.\left(\frac{\mathrm{d}A(0)}{\mathrm{d}C}-\frac{\mathrm{d}A(\boldsymbol{\kappa}-\boldsymbol{\tau})}{\mathrm{d}C}\right)\frac{1}{2B}\right]D(\boldsymbol{\kappa}-\boldsymbol{\tau})\right\}\times\left[1+\frac{I(\boldsymbol{\tau})-I(\boldsymbol{\kappa})}{B\overline{\nu}\,I^2(\boldsymbol{\kappa})}\right]^{-1},\tag{4.13}$$

if convoluted terms, which produce a dependence of moment on short range order are omitted.

$$a(0)=Ca_1(0)+(1-C)a_2(0)$$
$$A(\boldsymbol{\kappa}-\boldsymbol{\tau})=C^2A_1(\boldsymbol{\kappa}-\boldsymbol{\tau})+\left(1-C^2\right)A_2(\boldsymbol{\kappa}-\boldsymbol{\tau})$$
$$I(\boldsymbol{\kappa})=C^2I_1(\boldsymbol{\kappa})+2C(1-C)I_{12}(\boldsymbol{\kappa})+(1-C)^2I_2(\boldsymbol{\kappa})$$
$$B=CB_1+(1-C)B_2\tag{4.14}$$

are used in the above equations with the Fourier transforms

$$I(\boldsymbol{\kappa})=\sum_{R}I(\boldsymbol{R})\exp(\mathrm{i}\boldsymbol{\kappa}\cdot\boldsymbol{R})$$
$$A(\boldsymbol{\kappa}-\boldsymbol{\tau})=\sum_{R}a(\boldsymbol{R})\exp\{\mathrm{i}(\boldsymbol{\kappa}-\boldsymbol{\tau})\cdot\boldsymbol{R}\}\tag{4.15}$$

This expression reduces to the one magnetic species ferromagnetic case (eqn 3.39) for $\boldsymbol{\tau}=0$, $a_2(0)=A_2(0)=0$, $I_1=I_{12}=I_2$, and $B_1=B_2$. It also reduces to the one magnetic species antiferromagnetic case (Davis and Hicks 1981) for all except $\boldsymbol{\tau}=0$, and the two magnetic species ferromagnetic model just for $\boldsymbol{\tau}=0$.

The diffuse cross sections can be obtained from

$$\frac{\mathrm{d}\sigma}{\mathrm{d}\Omega}=\left(\frac{e^2\gamma}{2m_ec^2}\right)^2f^2(\boldsymbol{\kappa})\left|F_\perp(\boldsymbol{\kappa}-\boldsymbol{\tau})\right|^2$$
$$=\left(\frac{e^2\gamma}{2m_ec^2}\right)^2NC(1-C)S(\boldsymbol{\kappa}-\boldsymbol{\tau})f^2(\boldsymbol{\kappa})m_\perp(\boldsymbol{\kappa})m_\perp(-\boldsymbol{\kappa}),\tag{4.16}$$

where $m(\boldsymbol{\kappa})$ is defined by $F(\boldsymbol{\kappa} - \boldsymbol{\tau}) = m(\boldsymbol{\kappa})D(\boldsymbol{\kappa} - \boldsymbol{\tau})$, and $|D(\boldsymbol{\kappa} - \boldsymbol{\tau})|^2 = NC(1 - C)S(\boldsymbol{\kappa} - \boldsymbol{\tau})$. As with the ferromagnetic environment model parameters determined from the fit of the average staggered moment with composition can be used to predict the spatial moment fluctuations and the diffuse cross section.

4.3 γ-Mn–Ni ALLOYS

The face centred tetragonal form of manganese (γ-magnanese) can be stabilized at room temperature by the addition of many alloying elements. It is probably the simplest of the antiferromagnets in the 3d transition series of elements. The antiferromagnetic structure is shown in Fig. 4.4.
The alloys of γ-manganese with nickel are some of the most studied of antiferromagnetic metals. With little nickel the crystal structure is tetragonal below the Néel temperature, but as nickel is added the structure becomes cubic beyond about 15 at% after a narrow mixed phase region (Uchishiba 1971). It has recently been shown that the antiferromagnetic structure in the cubic phase is triple Q as shown in Fig. 4.5.

The most detailed investigation of the antiferromagnetic defect structure in the cubic phase of Mn–Ni is that of Moze and Hicks (1982). Polarization analysis was used to separate the purely magnetic scattering from the nuclear disorder scattering arising both from the distribution of the nickel in the manganese and from the distribution of the various nickel isotopes. The sample was a single crystal containing 27 at% nickel. Absolute values of the cross sections were obtained along three principal directions ⟨1 0 0⟩, ⟨1 1 0⟩, and ⟨1 1 1⟩. After correction for the isotopic disorder scattering from nickel and multiple scattering, the nuclear cross sections are shown in Fig. 4.6. The fitted lines are to Cowley short range order parameters which are listed in Table 4.1.

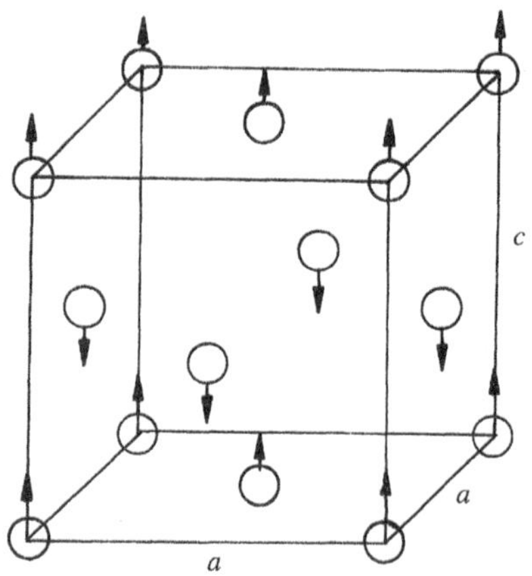

Fig. 4.4 Magnetic structure of the tetragonal form of γ-manganese alloys.

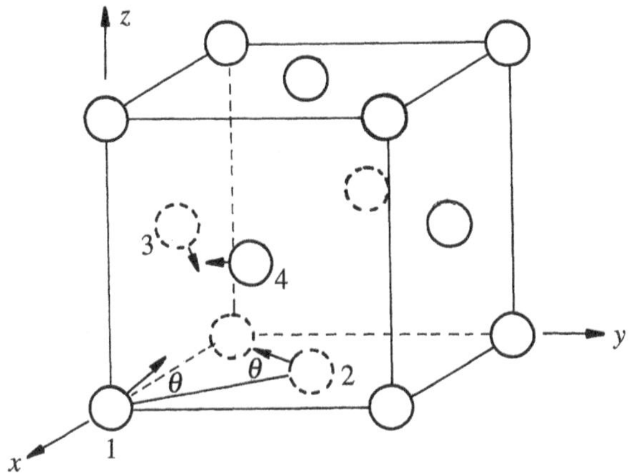

Fig. 4.5 The magnetic structure of cubic γ-manganese alloys.

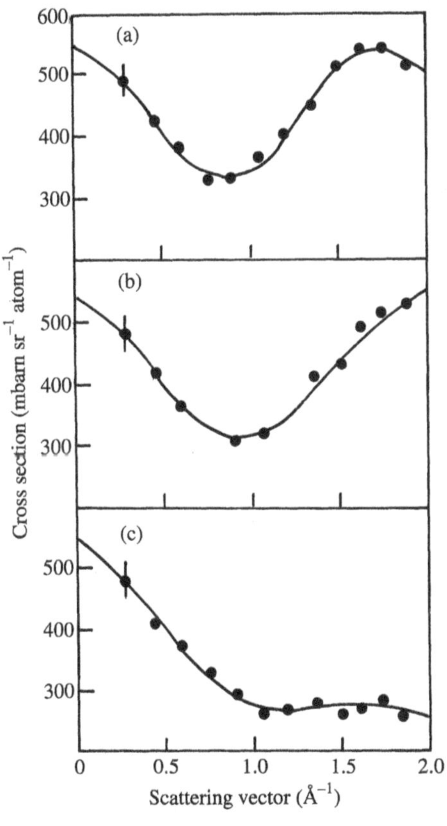

Fig. 4.6 Nuclear disorder scattering for $Mn_{73}Ni_{27}$ (after Moze and Hicks 1982) along (a) <1 0 0>; (b) <1 1 0> ; and (c) <1 1 1>. The solid lines are fitted to obtain the Cowley short range order parameters.

Table 4.1 Atomic short range order parameters for a $Mn_{73}Ni_{27}$crystal after Moze and Hicks (1982)

α_1	0.024(5)
α_2	0.058(5)
α_3	–0.010(5)
α_4	0.010(5)
α_5	0.001(5)
α_6	0.010(5)
α_7	–0.003(5)
α_8	–0.005(5)
α_9	0.004(5)
α_{10}	–0.003(5)

Table 4.2 Moment defect parameters for $Mn_{73}Ni_{27}$crystals after Moze and Hicks (1982)

$\bar{\mu}_{Ni}-\bar{\mu}_{Mn}$	–2.03 (10)μ_B
g'_1	–0.01(4)
g_1	1.62(4)
g_2	0.27(5)
g'_3	0.10(4)
g_3	–0.39(3)
g_4	0.00(1)
g'_5	–0.1(2)
g_5	–0.3(1)
$d\bar{\mu}/dC$	–3.0(3) atom^{-1}

The magnetic cross sections were fitted to a Marshall model similar to eqn (4.5) using

$$\frac{d\sigma}{d\Omega}=\left(\frac{e^2\gamma}{2m_ec^2}\right)^2 NC(1-C)S(\boldsymbol{\kappa}-\boldsymbol{\tau})f^2(\boldsymbol{\kappa})$$

$$\left\{\nu_{Ni\perp}-\nu_{Mn\perp}+(1-C)G_\perp(\boldsymbol{\kappa}-\boldsymbol{\tau})\right\}^2, \tag{4.17}$$

where the effect of short range order has been included but the dependence of the nickel moment on its environment and fluctuations of moment from fluctuations of short range order have been ignored. Moze and Hicks (1982) then assumed for simplicity that the antiferromagnetic structure was collinear and that the crystal consisted of three equally populated domains. The cross section for each crystal direction was averaged over the contributions from the three domains and the resulting expressions were simultaneously fitted to the data along the three crystal directions using the previously obtained parameters for the atomic short range order. The fit of the data to this form of the Marshall model is shown by the solid line in Fig. 4.7. The representation of the data is reasonable considering the experimental errors, but the value obtained for the variation of average staggered moment with composition from the fitted parameters was $d\nu/dC = -(6.5 \pm 0.4)\mu_B$ per added nickel atom compared with the measured $d\nu/dC = -(2.5 \pm 0.1)\mu_B$ atom^{-1} (Moze and Hicks 1982). The magnetic diffuse scattering is dominated by the peak at the {1 0 0} position which is most probably associated with the (0 0 1) point which is a reciprocal lattice point of the collinear antiferromagnetic structure. If so, any defect which is collinear with the structure will contribute little to the scattering at that point because the antiferromagnetic direction is parallel to the [0 0 1] direction. Elastic diffuse scattering

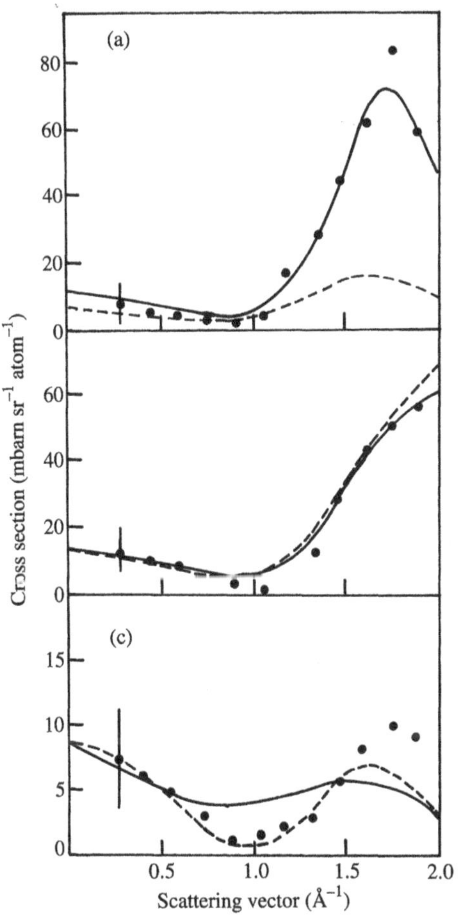

Fig. 4.7 Magnetic diffuse scattering for $Mn_{73}Ni_{27}$ (after Moze and Hicks 1982) along (a) <1 0 0>; (b) <1 1 0>; (c) <1 1 1>. The fits are described in the text.

associated with the (0 0 1) reciprocal lattice point must come from spatial fluctuations of moment transverse to the antiferromagnetic direction, and these are not covered by the Marshall model. The experiment was done at a low temperature and with low energy neutrons to minimize the contribution from magnon scattering. A fit excluding the points close to the peak in the diffuse cross section, and shown as a dashed line in Fig. 4.7, gave a much more reasonable value for $d\nu/dC = -(3.0 \pm 0.3)\mu_B$ atom^{-1}. The parameters for this fit are given in Table 4.2 where the two different defects for first and third neighbours refer to the sites on the same and opposite sublattices.

Moze and Hicks (1984) have also applied the magnetic environment model to the face centred antiferromagnetic Mn–Ni alloys. A fit of eqn (4.12) to the variation of average staggered moment with composition determined from magnetic Bragg scattering from powders (Moze *et al.* 1984) is shown in Fig. 4.8. Using the parameters obtained from the fit the magnetic diffuse cross sections were predicted in a manner similar to that for the ferromagnetic Ni–Pd alloys. Figure 4.9 shows the polycrystalline magnetic diffuse scattering cross sections for the range of antiferromagnetic alloys on which the solid lines are the prediction of the magnetic environment model. For most of the composition range the predictions fall well below the observed cross sections at scattering vectors near the {1 0 0} diffuse scattering peak. This is again true if the

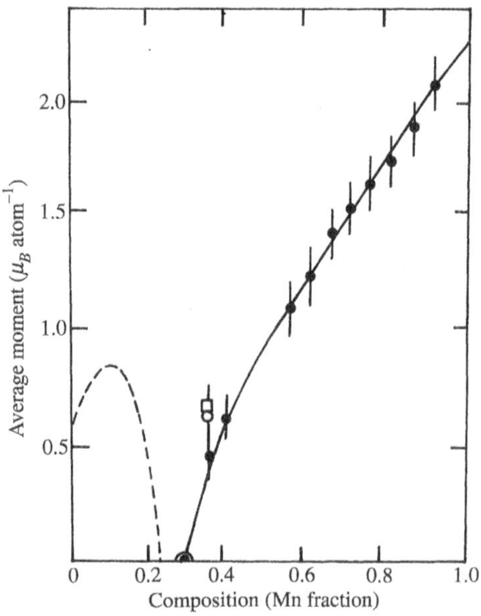

Fig. 4.8 Fit of the magnetic environment model to the variation of sublattice moment per atom with composition for Mn–Ni alloys. (After Moze and Hicks 1984.) The open points were from alloys not rotated and show the effects of preferred orientation.

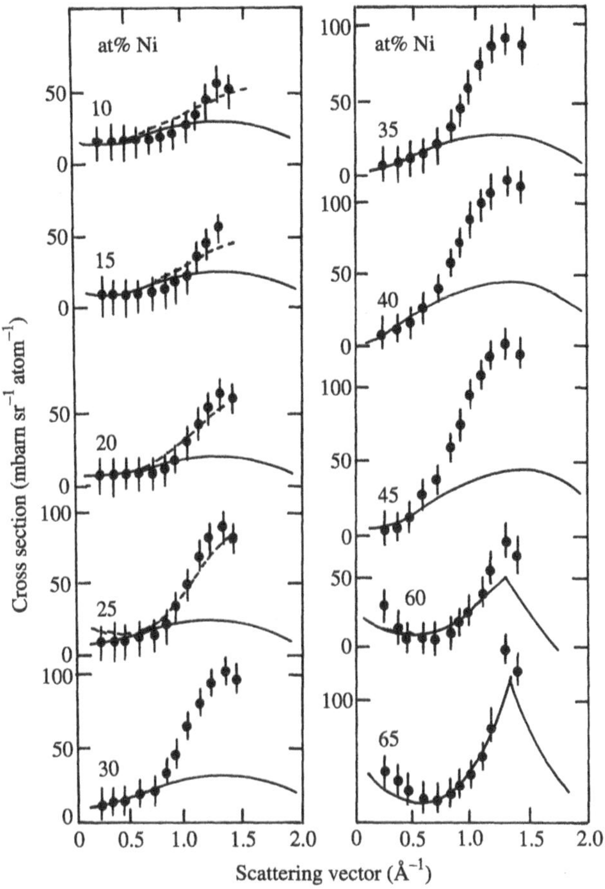

Fig. 4.9 Magnetic diffuse cross sections from polycrystalline Mn–Ni alloys covering the whole antiferromagnetic composition range. (After Moze and Hicks 1984.) The solid lines are the prediction of the collinear magnetic environment model.

predictions of the magnetic environment model are compared with the single crystal results of Moze and Hicks (1982). Figure 4.10 shows this comparison. Shown also as dashed lines are the fits of the Marshall model ignoring the points shown as open. It is clear that the magnetic environment model and the Marshall model have a problem with the scattering near {100} most probably because there is no provision in either for defects which are not collinear with the magnetic direction.

In retrospect the noncollinear structure shown in Fig. 4.5 should have been used for the analysis of the magnetic diffuse scattering from Mn–Ni alloys. However, in the absence of single domain samples the structures have to be distinguished by non-neutron means (Kennedy and Hicks 1987; Kawarazaki *et al.*

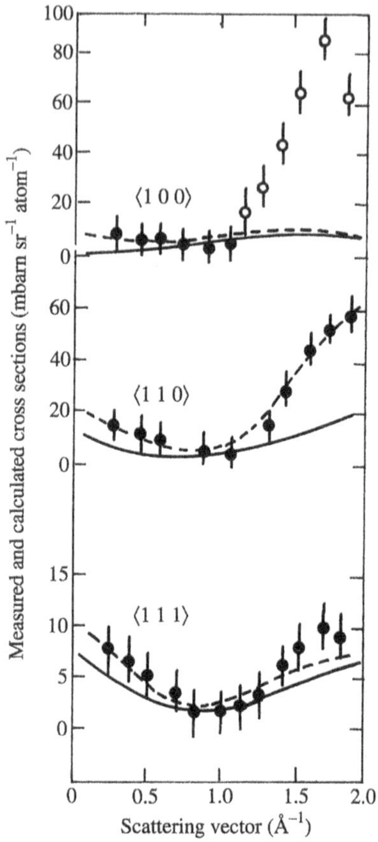

Fig. 4.10 Magnetic diffuse scattering from the single crystal of $Mn_{73}Ni_{27}$. The magnetic environment model predictions (full lines) and the Marshall model fits (dashed lines). (After Moze and Hicks 1984.)

1988). Rainford (private communication) has developed a variation of the Marshall model for the triple Q structure of Mn–Ni in which the moment change at a particular site is the vector sum of a moment change parallel to the direction of its sublattice and a moment change parallel to the direction of the sublattice on which the defect is situated. This means that moments are rotated out of the direction of their sublattice and may change their magnitude. For any set of neighbours in which the atoms are not on the same sublattice as the defect, two parameters describe the moment changes. This can be compared with the collinear case where neighbours not all on the same sublattice as the defect need two parameters for their moment change, one for one sublattice and one for the other. Two parameters are also needed for the triple Q structure and are linearly related to those for the collinear structure. It can be shown that Rainford's variation of the Marshall model yields exactly the same form for the cross section as does the collinear Marshall model when the sum over the three domains is taken, and suffers from the same inability to describe the scattering near {1 0 0} with realistic values of $d\nu/dC$. In fact, fits using both models yield the same values for $d\nu/dC$.

The Mn–Fe alloy system, at least in the middle composition range, has similar features to Mn–Ni. There is again a relatively sharp diffuse peak near the {100} position (Hicks 1988) which cannot easily be fitted using a collinear model for the magnetic defects (Kennedy and Hicks 1989).

4.4 ANTIFERROMAGNETIC Mn–Cu ALLOYS

In contrast to the Mn–Ni system all the antiferromagnetic Mn–Cu alloys are tetragonal and the magnetic structure is collinear (Meneghetti and Sidhu 1957). The diffuse {1 0 0} peak was first seen by Bacon *et al.* (1957) who associated it with the (0 0 1) and therefore with transverse moment deviations, by cooling a polycrystalline alloy through the Néel temperature under a compressive stress. This had the effect of aligning the antiferromagnetic domains and tetragonal axes along the nearest [0 0 1] direction and thus giving the sample an overall preferred orientation. It was noted that the {1 0 0} diffuse feature was enhanced with the stress direction along the scattering vector.

A similar experiment was performed by Cywinski *et al.* (1980) but incorporating polarization analysis. The results are shown for stress direction parallel to the scattering vector in Fig. 4.11, and for stress direction perpendicular to the scattering vector in Fig. 4.12. A clear separation is seen between magnetic and nuclear scattering, and in particular that the {1 0 0} feature at $\kappa = 1.75$ Å^{-1} is magnetic. What is also obvious is that, at least for alloys containing more than 80 at% manganese, the {1 0 0} feature is associated with the (0 0 1) reciprocal lattice position and for well oriented alloys is relatively sharp. For alloys containing less than 80 at%manganese the *c*/*a* ratio is closer to one and not so much preferred orientation is induced. Detailed analysis of these results is precluded because of the partial nature of the preferred orientation and the extreme clustering of the atoms seen in the large small $\boldsymbol{\kappa}$ nuclear cross sections.

The same feature at the (0 0 1) position is seen in experiments with single crystals of Mn–Cu. Tsunoda and Nakai (1980) mapped the elastic intensity at the (0 0 1) position and initially identified it as a Bragg peak because of its sharpness. This then led them to suggest that the antiferromagnetic structure was canted away from the [0 0 1] direction. From the above discussion it is clear that the intensity is not Bragg in origin, because this would lead to a sharp feature in polycrystalline diffraction patterns which is not observed. The extremely sharp nature of this intensity was further reported by Hicks *et al.* (1988), from a polarization analysis experiment on an almost single domain 90 at % manganese single crystal, which again showed its magnetic character. In this case the explanation given for the existence of the peak involved a long range canting of the magnetic environment around each copper defect. So that no Bragg scattering

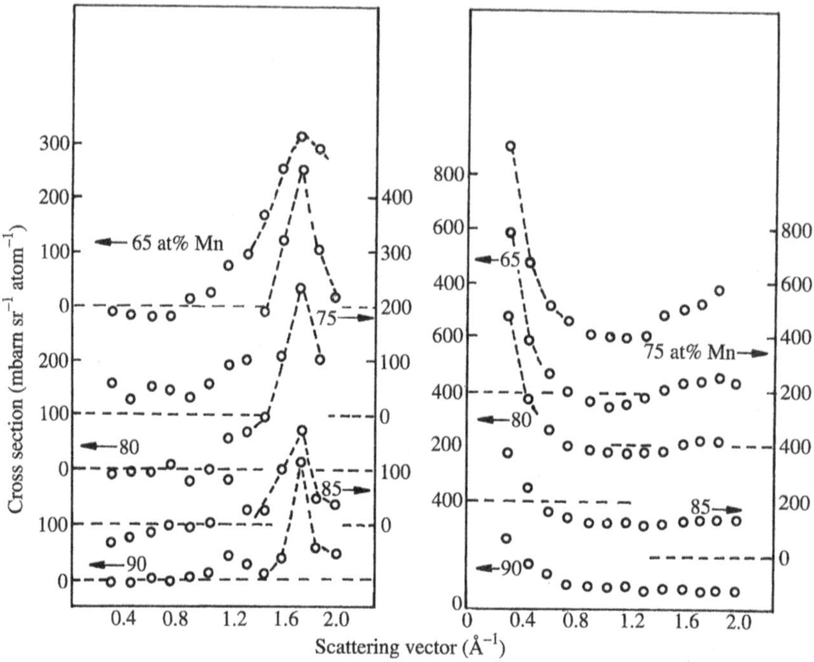

Fig. 4.11 The diffuse magnetic and nuclear defect cross sections for stressed γ-Mn–Cu alloys at 4.2 K with the stress parallel to κ. (After Cywinski *et al.* 1980.)

was generated by the canting, the direction of the cant around each defect was assumed to be different. Whatever the final form for the origin of the (0 0 1) magnetic peak in these alloys it appears that it will involve some deviation of moment from the antiferromagnetic direction. If Rainford's model for the triple Q structure is applied to the collinear structure his two parameters for the moment deviation (a component parallel to the site sublattice and a component parallel to the defect site sublattice) will both be along the antiferromagnetic direction. So it appears that more than this is required to produce moment deviations transverse to the antiferromagnetic direction. This then implies that something more is required also for alloys with the triple Q structure.

4.5 NONMETALLIC ANTIFERROMAGNETIC ALLOYS

In many ways nonmetallic magnetic materials are simpler than metallic materials. The localization of the magnetic electrons means that the magnitude of the moment at each site is unaffected by its environment. This leaves the broken coupling between defect and surroundings as the sole determinant of the wider defect. Lovesey and Marshall (1966) have developed a linearized molecular field model for systems in which the impurity is nonmagnetic. The

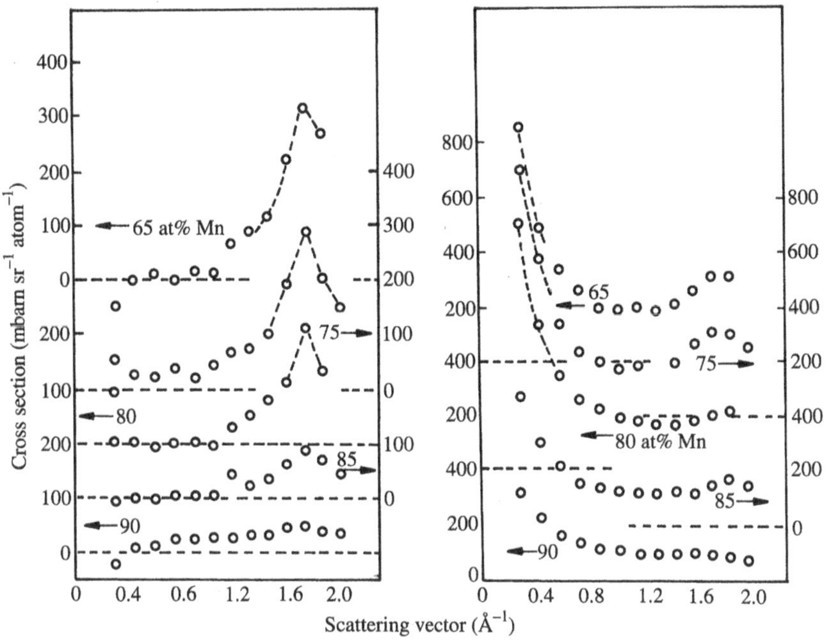

Fig. 4.12 The diffuse magnetic and nuclear defect cross sections for stressed γ-Mn–Cu alloys at 4.2 K with the stress perpendicular to κ. (After Cywinski *et al.* 1980.)

model predicts a moment change primarily on first neighbours in the cubic lattices which increases with temperature.

There is one test of the Lovesey and Marshall (1966) model, that of Bakshi and Hicks (1984), which used a $Mn_{0.68}Zn_{0.32}F_2$ single crystal and polarization analysis to isolate the magnetic defect cross section. The test is not ideal because the model is for dilute impurities only and for cubic not tetragonal lattices. The latter is not a serious problem because the antiferromagnetic interaction between body centred and corner atoms in the tetragonal cell of $Mn_{1-C}Zn_CF_2$ is by far the strongest interaction, so that the coordination is similar to body centred cubic.

The experiment is difficult because of considerable magnetic inelastic scattering. By using a time of flight technique in which the flipper in a polarization analysis experiment was rapidly turned on and off by a pseudorandom sequence, the difference between elastic spin flip and non spin flip cross sections could be isolated. The elastic magnetic diffuse cross section was then extracted by assuming that the nuclear defect scattering was isotropic [no short range order (Shull *et al.* 1951)], and by inserting known values for the isotopic disorder and the nuclear spin incoherent cross sections. Points resulting from this procedure are shown in Fig. 4.13 for the [1 0 0] direction and in Fig. 4.14 for the [1 1 0] direction show a definite modulation of the diffuse cross section. The solid lines are a fit of the equations

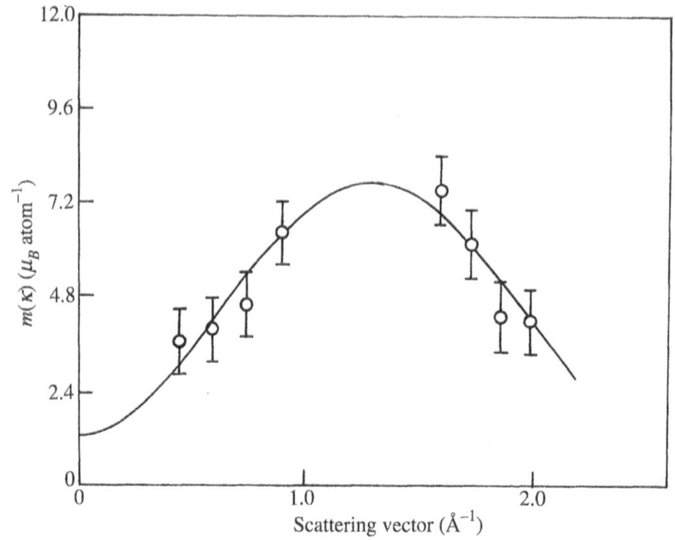

Fig. 4.13 The variation of $m(\kappa)$ at 30 K in the [1 0 0] direction for $Mn_{0.68}Zn_{0.32}F_2$. The full line is a fit of the Marshall model including a lattice distortion effect. (After Bakshi and Hicks 1984.)

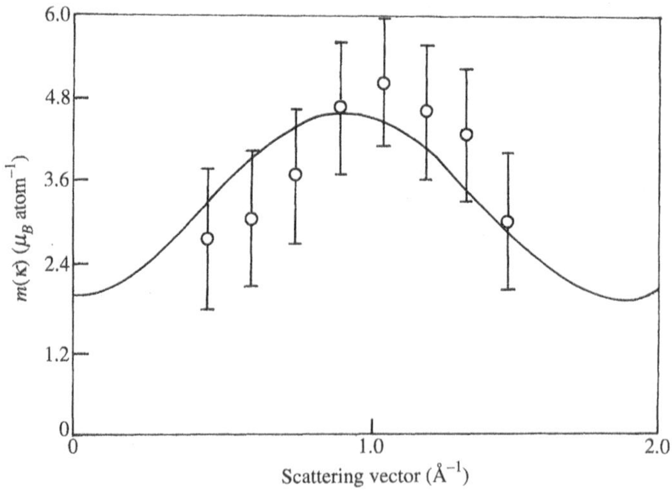

Fig. 4.14 The variation of $m(\kappa)$ along the [1 1 0] direction for $Mn_{0.68}Zn_{0.32}F_2$. The full line is a fit of the Marshall model without the lattice distortion effect. (After Bakshi and Hicks 1984.)

$$\frac{d\sigma}{d\Omega} = \left(\frac{e^2\gamma}{2m_e c^2}\right)^2 NC(1-C)f^2(\boldsymbol{\kappa})\big[-\bar{\mu}_{\mathrm{Mn}\perp} + 2(1-C)\,g_\perp(\boldsymbol{R}_1) + 8(1-C)\,g_\perp(R_2)\cos(\kappa\alpha/2)\big]$$

for the [1 0 0] direction, and (4.18)

$$\frac{d\sigma}{d\Omega} = \left(\frac{e^2\gamma}{2m_e c^2}\right)^2 NC(1-C)f^2(\boldsymbol{\kappa})\big[-\bar{\mu}_{\mathrm{Mn}\perp} + 2(1-C)g(R_1) + 4(1-C)g_\perp(R_2) + 4(1-C)g_\perp(R_2)\cos\left(\kappa\alpha/\sqrt{2}\right)\big]$$

for the [1 1 0] direction from the Marshall model [eqn 4.5]. It is clear from these equations that, for the directions measured, it is impossible to determine the average manganese moment and the defect on the first neighbours independently. The first neighbours in this body centred tetragonal cell are the adjacent body centres along the [0 0 1] direction. So the comparison with Lovesey and Marshall (1966) must be with the second neighbbour defect, and this is shown in Fig. 4.15. The second neigbours are at the corners of the cell and are strongly antiferromagnetically coupled so we would expect that they would be strongly affected by the elimination of this coupling with the body centred atom and this is clear from the results. The solid line is the prediction of Lovesey and Marshall (1966) and it lies significantly below the experimental points. This is expected as the model is for dilute systems and no consideration has been taken of defect

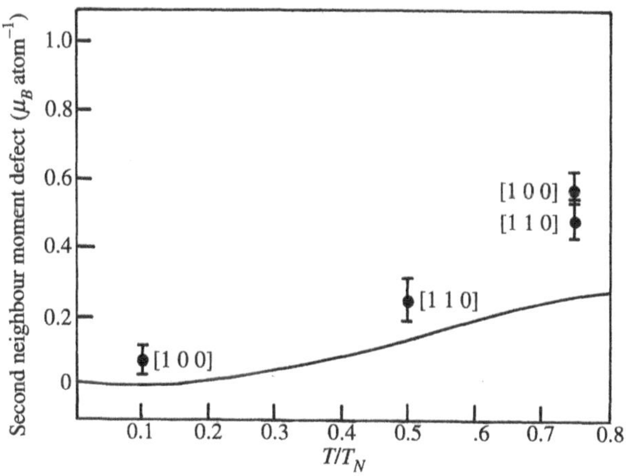

Fig. 4.15 The reduction of the ordered magnetic moment on the second neighbour in $Mn_{0.68}Zn_{0.32}F_2$ as a function of temperature. The full line is the theoretical result for dilute concentrations of a nonmagnetic impurity in a Heisenberg magnet. (After Bakshi and Hicks 1984.)

overlap, and also molecular field models always overestimate the degree of order. Similar results (Bakshi and Hicks 1983) were obtained using powder samples of $Mn_{1-c}Zn_cF_2$. The first neighbour defect determined from these results was close to zero, which is in line with the very small interaction between first neighbours.

The study of defects in antiferromagnetic materials is still in its infancy. The models which could provide a framework for the analysis of the neutron scattering from collinear antiferromagnets have been presented in this chapter but it has been largely used to show that the effects observed cannot be described in this way. This has been so because the data from the γ-manganese based alloys have been dominated by noncollinear effects and these will have to be described before any progress is made in the application of the collinear models and the consequent characterization of the electronic processes in the γ-manganese alloys.

References

Bacon, G.E., Dunmur, I.W., Smith, J.H., and Street, R. (1957). *Proc. Roy. Soc.* **A241**, 223.

Bakshi, E. and Hicks, T.J. (1983). *J. Phys. C* **16**, 2337.

Bakshi, E. and Hicks, T.J. (1984). *J. Phys. C* **17**, 1153.

Cywinski, R., Wells, P., Campbell, S.J., and Hicks, T.J. (1980). *Nuk.* **25**, 787.

Davis, J.R. and Hicks, T.J. (1981). *Solid State Comm.* **39**, 1223.

Davis, J.R., Cywinski, R., Moze, O., and Hicks, T.J. (1982). *J. de Phys.* **43**, C7-65.

Hicks, T.J. (1988). *Mat. Sc. Forum* **27-28**, 167

Hicks, T.J., Moze, O., and Rainford, B.D. (1988). *J. de Phys.* **49**, C8-163.

Kawarazaki, S., Fujita, K., Yasuda, K., Sasaki, Y., Mizusaki, T., and Hirai, A. (1988). *Phys. Rev. Lett.* **61**, 471.

Kennedy, S.J. and Hicks, T.J. (1987). *J. Phys. F* **17**, 1599.

Kennedy, S.J. and Hicks, T.J. (1989). *J. Mag. Mag. Mat.* **81**, 56.

Lovesey, S.W. and Marshall, W. (1966). *Proc. Phys. Soc.* **89**, 613.

Maier, B. (ed.) (1983). *Neutron research facilities at the ILL high flux reactor.* Institut Laue–Langevin, Grenoble.

Marshall, W. (1968). *J. Phys. C* **1**, 88.

Meneghetti, D. and Sidhu, S.S. (1957). *Phys. Rev.* **105**, 130.

Moze, O. and Hicks, T.J. (1982). *J. Phys. F* **12**, 1.

Moze, O. and Hicks, T.J. (1984). *J. Phys. F* **14**, 221.

Moze, O., Hicks, T.J., and Blanckenhagen, P.von (1984). *J. Mag. Mag. Mat.* **42**, 103.

Schärpf, O. (1982). Recent advances with supermirror polarisers. In *Neutron scattering—1981*, (AIP Conference Proceedings No.89), (ed. John Faber Jr,) p. 182. American Institute of Physics, New York.

Shull, C.G., Strauser, W.A., and Wollan, E.O. (1951). *Phys. Rev.* **83**, 333.

Tsunoda, Y. and Nakai, Y. (1980). *Solid State Comm.* **34**, 413.

Uchishiba, H. (1971). *J. Phys. Soc. Jap.* **31**, 436.

5

Excitations

Atomic randomness makes a qualitative difference to the excitations in materials with long range magnetic order. Even to predict the correct number of modes, full account must be taken of the range of atomic environments for most materials. This is quite difficult and has only been done for a limited number of materials within the coherent potential approximation. Excitations near the critical temperature for some types of random systems can show very complicated behaviour even to time-dependent effects measurable on laboratory time scales. This chapter explores some of the experiments in which the excitations in magnetically ordered alloys have been measured and the diverse constructs for their analysis.

5.1 AVERAGE MAGNONS IN FERROMAGNETIC ALLOYS

The simplest way to characterize the elementary excitations in magnetically ordered alloys is to ignore the differences between exchange interactions and moments and replace the individual interactions and moments by their averages. Clearly this approach only has a chance of yielding reasonable results for low q magnons where the long wavelength can average over the spatial inhomogeneity of the alloy.

For a Heisenberg system the Hamiltonian including just exchange can be written

$$\mathcal{H} = -2\sum_{i<j} j(\boldsymbol{r}_i - \boldsymbol{r}_j)\boldsymbol{S}_i \cdot \boldsymbol{S}_j. \tag{5.1}$$

If we define the spatial moments of J for a cubic lattice as

$$j^{(n)} = z^{-1}\sum_{r} j(\boldsymbol{r})\left(\frac{r}{\boldsymbol{R}_{\mathrm{n}}}\right)^{n} \tag{5.2}$$

where z is the number of first neighbours and R_{n} the first neighbour distance. The Hamiltonian gives the ground state energy as

$$E = -Nz\,j^{(0)}S^2 \tag{5.3}$$

and the dispersion relation as

$$\hbar\,\omega = Dq^2 + O(q^4) \tag{5.4}$$

for long wavelength magnons with wavevector $\boldsymbol{q}$. At low temperatures

$$D = \tfrac{1}{3} z \, \boldsymbol{R}_n^2 j^{(2)} S. \tag{5.5}$$

To apply the above to a binary alloy three functions $j_{XY}(r)$, where $X,Y \equiv A,B$, need to be introduced and an effective exchange interaction can be defined by

$$j_{\text{eff}}^{(n)} \overline{S^2} = C_A^2 j_{AA}^{(n)} S_A^2 + 2C_A C_B j_{AB}^{(n)} S_A S_B + C_B^2 j_{BB}^{(n)} S_B^2 \tag{5.6}$$

which gives the ground state energy as

$$E = -Nz \, j_{\text{eff}}^{(\text{o})} \overline{S^2} \tag{5.7}$$

and coefficient of the q^2 term of the low temperature magnon energy as

$$D = \tfrac{1}{3} z \, \boldsymbol{R}_n^2 j_{\text{eff}}^{(2)} \overline{S}. \tag{5.8}$$

Most ferromagnetic materials are metallic and this adds a further level of approximation, in that the Heisenberg Hamiltonian is inappropriate for metals. Nevertheless, several studies have been able to use the average magnon model to meaningfully interpret their data. Hatherley *et al.* (1964) carried out an extensive survey of the magnon stiffness constant D in ferromagnetic Fe–Ni alloys both in the b.c.c and f.c.c forms. Their novel experimental technique, using polycrystalline samples, enabled the measurement of D in situations where single crystals might not be obtainable, and they were able to obtain measurements on the same sample in the b.c.c. and f.c.c. forms. For a magnon dispersion relationship proportional to q^2 the magnon scattering is confined within $\theta_c = E/k^2 D$ of the forward direction, where E is the neutron energy and k is the neutron wavevector. E/k^2 is constant for all neutron energies and the experiment uses a white neutron beam. Measurement of θ_c yields magnon stiffness D.

Figure (5.1) shows the values of D obtained by Hatherley *et al.* (1964) and also the values of $j_{\text{eff}}^{(n)}$ derived by application of eqn (5.8) knowing $\overline{S}$. To check whether the application of the average theory was making sense Hatherley *et al.* (1964) compared their effective exchange interactions with a crude estimate from the Curie temperatures across the composition range also shown in Fig. 5.1. The values of $j_{\text{eff}}^{(n)}$ were then used to extract the three individual exchange interactions between the two species for the two crystal structures. These are shown in Table 5.1.

It appears that the model is largely successful in that the variation of $j_{\text{eff}}^{(2)}$ with composition and across two phases compares well with the variation of the Curie temperature corrected for average spin and coordination. Additionally, the analysis reveals that in the face centred cubic phase the Fe–Fe first neighbour interaction is antiferromagnetic which accords with the observed antiferromagnetic structure of f.c.c. iron (Abrahams *et al.* 1962). In spite of the obvious attractiveness of the approach a simple rigid band model of the Fe–Ni alloys system fails to adequately account for the variation of magnon stiffness with composition (Lowde *et al.* 1965).

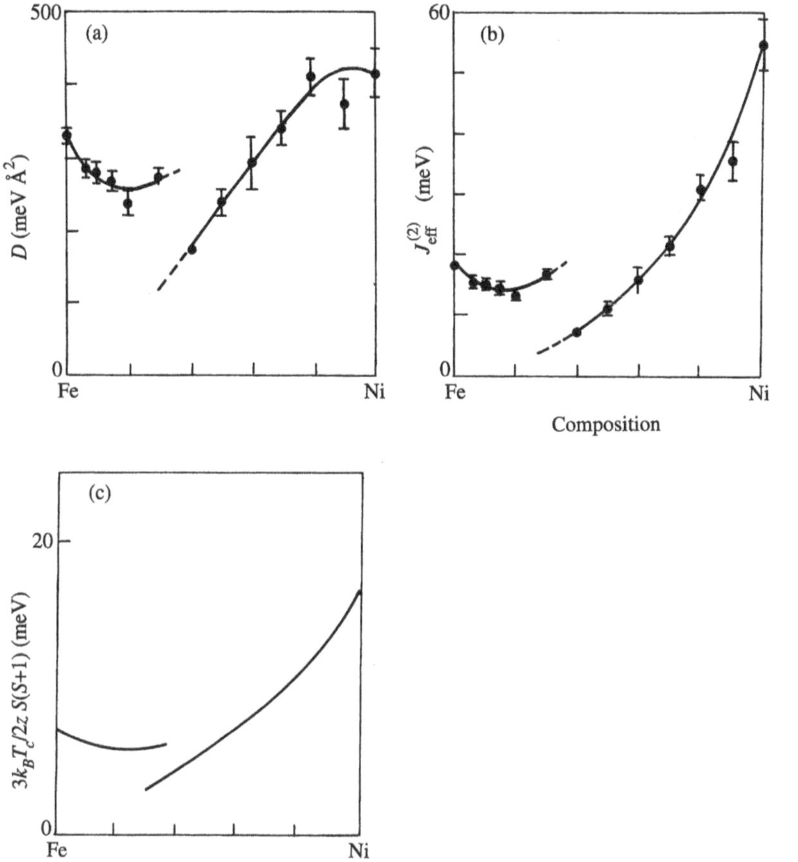

Fig. 5.1 (a) D, the observed coefficient of q^2 in the magnon energy, and (b) the second moment of $J(r_{ij})$ for Fe–Ni alloys as a function of composition. (c) An estimate of the exchange interaction from the Curie temperatures. (After Hatherley *et al.* 1964.)

Table 5.1 Second moments of $j_{XY}(r)$ in Fe–Ni alloys (meV)

	Body centred	Face centred
$j^{(2)}_{\text{FeFe}}$	18.2±0.5	−9±2.6
$j^{(2)}_{\text{FeNi}}$	−41±13	39±5
$j^{(2)}_{\text{NiNi}}$		52±3

Measurements on another ferromagnetic alloy system, Ni–Mn (Hennion *et al.* 1977), also give a qualitatively correct picture when analysed in the same average Heisenberg manner. Hennion *et al.* (1977) used single crystals and more conventional triple axis methods to obtain the magnon stiffness. The first neighbour interactions coming from the analysis are positive for Ni–Ni and Ni–Mn pairs but negative for Mn–Mn pairs. This then accords with the fact that f.c.c. manganese is antiferromagnetic.

Hennion *et al.* (1977) also observed that the magnon widths increased considerably with both composition and magnon energy. This is a reminder that it is not sufficient to describe these excitations using an average model. There are significant deviations from the average which increase with composition and also with magnon wavevectors as the magnons sample the more local environment.

5.2 THE CLUSTER MODEL FOR ANTIFERROMAGNETIC INSULATORS

One of the simplest ways to demonstrate and to estimate the amount of broadening of the magnon response function is by using the calculated frequency distribution of different individual atom clusters which occur in the alloy system. This can be done for binary insulating systems, normally called mixed crystals, and gives a good estimate of the spectral response function near the zone boundary where the excitations are representative of spatially local regions.

Even for dilute alloys where an impurity is completely surrounded by the host there are three different environments for a system where first neighbour interactions dominate. The first is for a host atom surrounded by other host atoms, the second is for an impurity atom surrounded by host atoms, and the third is for host atoms who have a first neighbour impurity. For more concentrated alloys there is a wider spectrum of environments. Consider an AB alloy in which there is a concentration C of S_A, and $(1 - C)$ of S_B which interact only with first neighbour exchange constants j_A, j_B, and j_{AB}. The probability that an atom has r A neighbours and $z - r$ B neighbours is

$$p\begin{pmatrix} r \\ z \end{pmatrix} = \frac{z!}{r!(z-r)} C^r (1-C)^{z-r}. \tag{5.9}$$

For an A spin, the excitation frequency within molecular field theory is

$$rj_A S_A + (z-r) j_{AB} S_B \tag{5.10}$$

and for a B spin is

$$rj_{AB} S_B + (z-r) j_B S_B. \tag{5.11}$$

The scattering cross section may then be calculated as a series of delta functions at these excitation frequencies with relative intensities given by

$$Cg^2 \mu_B^2 S_A^2 p\left(\begin{smallmatrix} r \\ z \end{smallmatrix}\right) \tag{5.12}$$

for the A sites, and

$$(1-C) g^2 \mu_B^2 S_B^2 p\left(\begin{smallmatrix} r \\ z \end{smallmatrix}\right) \tag{5.13}$$

for the B sites. j_A and j_B are known from the pure materials and for insulators the predominant interaction is usually the first neighbour one. j_{AB} can be obtained from optical experiments on dilute concentrations of one species in the other where one species is completely surrounded by atoms of the other or it may be calculated (Cowley and Buyers 1972).

An early and good example of the aptness of this approach is the (Co,Mn)F_2 system (Buyers *et al.* 1971). Figure 5.2 shows magnon dispersion curves obtained for $Co_{0.7}Mn_{0.3}F_2$. There are two relatively well-defined modes which are significantly different to the modes in the pure materials. Each mode, at the zone boundary, is associated with the spectrum of the two magnetic species in their respective range of atomic environments. Figure 5.3 shows the neutron scattering intensity at the zone boundary and the shape of the response expected from the cluster model. The solid lines actually join the intensities predicted for the delta function modes resulting from the frequencies associated with the various clusters. It is particularly noteworthy that the difference in widths for the two modes is reproduced by the cluster model.

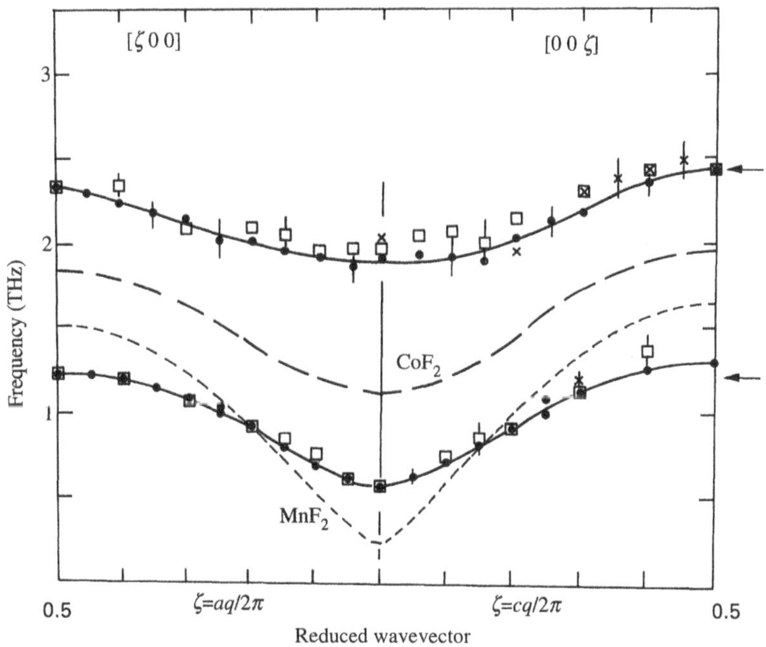

Fig. 5.2 Dispersion relations for magnetic excitations propagating along the *a* and *c* directions in $Co_{0.7}Mn_{0.3}F_2$. (●) (0 0 1) zone (magnetic), (×) (1 0 0) zone magnetic, (□) (1 0 1) zone nuclear. The pure compounds are shown for comparison. Arrows are the predictions of molecular field theory for the cluster model. (After Buyers *et al.* 1971.)

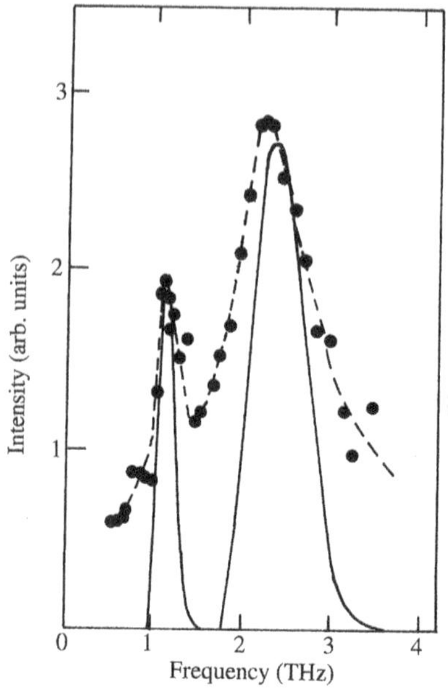

Fig. 5.3 Neutron scattering at the boundaries in $Co_{0.7}Mn_{0.3}F_2$ ($[\zeta\ 0\ 1]$, $\zeta = 0.5$). The experimental results are compared with an Ising cluster model with smooth curves drawn through the intensities of the discrete frequencies. (After Cowley and Buyers 1972.)

5.3 MODES AT GENERAL WAVEVECTORS

It is clear from the previous section that at least at the zone boundary there are many modes arising from the range of environments experienced by the two species. In general these will by grouped into two excitation bands each associated with one of the two species in the alloy as is the case for $(Mn,Co)F_2$. The average magnon model, however, makes no provision for more than one mode. The question is whether this is simply a function of the different wavevector ranges of applicability of the two models, or whether there are circumstances where at least two modes exist over the entire wavevector range. The answer is that there are circumstances where the average magnon does not work even at small wavevectors. This is when, for instance, the single site excitations of an impurity moment lie at quite a different energy to the magnon band of the host. The two vibrations can then be regarded as quite separate. This is the case for cobalt impurities in $KMnF_3$ as shown in Figure 5.4 (Buyers *et al.* 1973) where even for 20 per cent of cobalt the local cobalt excitation mode is quite separate from the propagating magnons whose dispersion is very little altered from pure $KMnF_3$. If the energies of the excitations were not so separate they would obviously mix and the situation would be much more complex. To make further progress we need to review the application of the coherent potential approximation to the problem of magnetic excitations in these alloy systems.

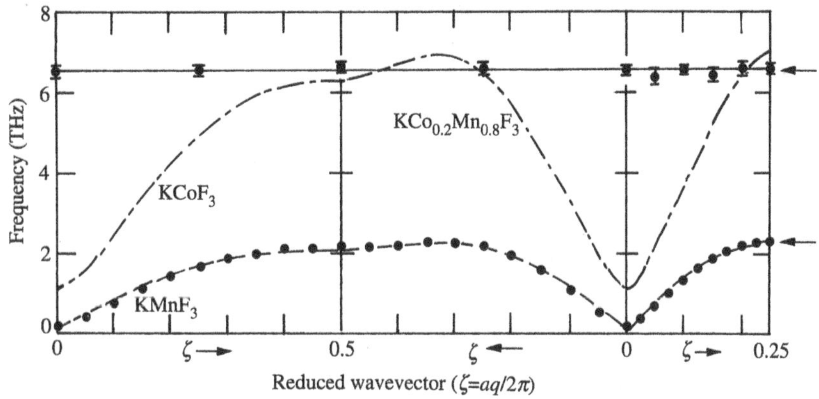

Fig. 5.4 The dispersion relations for magnons in $KCo_{0.2}Mn_{0.8}F_3$. Arrows on the right indicate predictions of a cluster molecular field theory. (After Buyers *et al.* 1971.)

From eqn (1.37) the cross section for inelastic scattering is

$$\left(\frac{d^2\mathbf{s}}{d\Omega dE}\right)_{k,k'} = \frac{V}{\pi}\frac{k'}{k}\left(\frac{e\gamma}{\hbar c}\right)^2 \sum_{\mathbf{a}} \frac{\chi''_{\mathbf{aa}}(\boldsymbol{\kappa},\omega)}{\left(1-e^{-\mathbf{b}\hbar\omega}\right)}. \tag{5.14}$$

The susceptibility can be related to a Green function, G,

$$\chi''_{\alpha\alpha}(\omega) = \frac{1}{2i}\frac{g^2}{V}\left[G(\omega+i\epsilon) - G(\omega\ -i\epsilon)\right] \tag{5.15}$$

(Buyers *et al.* 1971; Cowley and Buyers 1972) and the theory calculates the Green function for the mixed system within the CPA.

It is worth having a brief look at the detailed analysis of Buyers *et al.* (1971) which is applied to the $(Mn,Co)F_2$ and $K(Mn,Co)F_3$ mixed antiferromagnetic systems. As well as including different potentials on sites occupied by the two magnetic species, Buyers *et al.* (1971) allow for the the different occupations of the immediate environment of each species affected directly by the exchange coupling. This is done taking into account the probabilities of the various environments in a manner similar to eqn (5.9). All the potentials weighted by their probabilities are then combined to extract an effective site potential $\Sigma(\omega)$ by requiring in the CPA manner that replacement of the effective potential at each site by the real potential make no difference to the average scattering of magnons.

The configurationally averaged Green's function for the alloy is given by

$$\langle G(\boldsymbol{\kappa},\omega)\rangle_c = \frac{1}{D(\boldsymbol{\kappa},\omega)}\begin{pmatrix} \omega + A + \Sigma(-\omega) & -B\gamma(\boldsymbol{\kappa}) \\ B\gamma(\boldsymbol{\kappa}) & \omega - A - \Sigma(\omega) \end{pmatrix} \tag{5.16}$$

with A and B constants which are optimized along with Σ and

$$\gamma(\boldsymbol{\kappa}) = z^{-1}\sum_{\Delta} e^{i\boldsymbol{\kappa}\cdot\Delta} \tag{5.17}$$

where the Δ are the vectors linking an atom to its neighbours. The denominator

$$D(\boldsymbol{\kappa},\omega) = \{\omega + A + \Sigma(-\omega)\}\{\omega - A - \Sigma(\omega)\} + B^2\gamma^2(\boldsymbol{\kappa}) \tag{5.18}$$

largely determines the resonances in the response and therefore the energies of the maxima in the magnon cross section. Figure 5.5 illustrates how essentially two resonances occur in the magnon cross section at the zone boundary in $Mn_{0.3}Co_{0.7}F_2$. At the zone boundary $\gamma(\boldsymbol{\kappa})$ is zero and the roots of the denominator of the Green function are

$$\pm\omega = A + \Sigma(\pm\omega). \tag{5.19}$$

So in addition to the computed variations of the real and imaginary parts of $\Sigma(\omega)$ the straight line on the figure is $A - \omega$. It can be seen that the line intersects the real part of Σ at two points which will correspond approximately to the resonances in the Green function. The computed variation of the dissipative part of the the susceptibility ($\chi''(\omega)$) shows those peaks with the structure in the left-hand peak presumably due to the structure in the two parts of Σ in this frequency range. What can also be seen is that, depending on the form of $\Sigma(\omega)$, there need only be one intersection of the line with Σ and therefore only one resonance in the Green function. The CPA calculation therefore confirms the essential correctness of the cluster model at the zone boundary. However, the CPA is an approximation which is good over the whole of the zone and can be used to calculate the intensity of the neutron scattering from magnetic excitations at any point. Figure 5.6 shows the results of the CPA calculations of neutron intensity

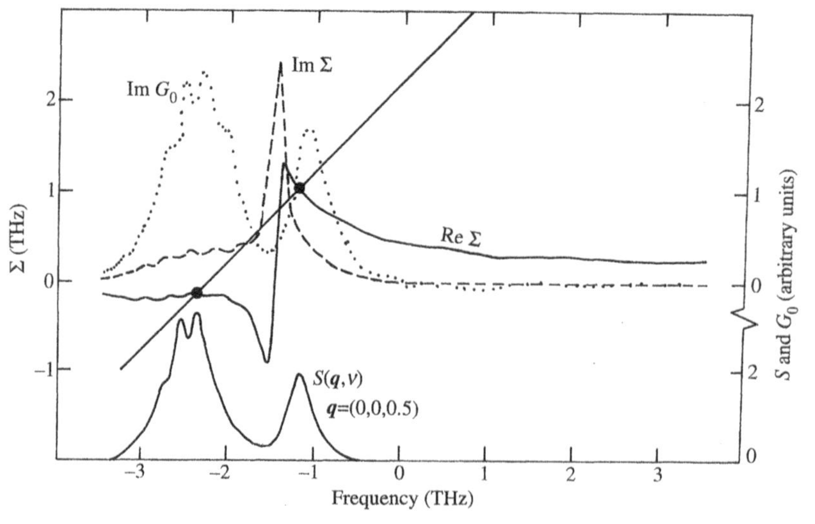

Fig. 5.5 The real and imaginary parts of the averaged potential $\Sigma(\omega)$ and the dissipative part of the susceptibility $\chi''(\boldsymbol{\kappa},\omega)$ in $Mn_{0.3}Co_{0.7}F_2$. The straight line is $A-\omega$. (After Buyers *et al.* 1972.)

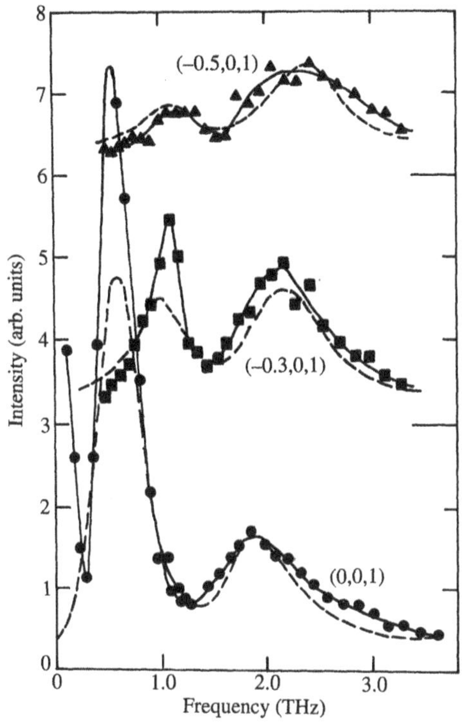

Fig. 5.6 Neutron scattering lineshapes in $Mn_{0.3}Co_{0.7}F_2$ (points and solid lines) compared with the CPA calculations folded with the instrumental resolution. (After Buyers *et al.* 1972.)

at other points in the zone. As can be seen the agreement with the position, shape, and width of the resonances is good.

Paradoxically, the Green function–CPA treatment of the magnetic excitations has more problems for mixed antiferromagnetic systems in which one species has zero moment. This is due firstly to the normalization of the Green function and can be solved by putting a spin on the nonmagnetic impurity but leaving it uncoupled to the rest of the antiferromagnetic order. However, this leads to problems with CPA in which the magnons are allowed to propagate on the nonmagnetic sites. Buyers *et al.* (1972) excluded the magnons from the nonmagnetic sites by coupling the spins on those sites very strongly. This leads to a high energy localized mode which could be recognized and excluded from the results. By these means Buyers *et al.* (1972) were able to well characterize the modes in $(Mn,Zn)F_2$. Figure 5.7 shows the calculated and measured magnon lineshapes at two positions in the Brillouin zone. There is similar agreement when the CPA is applied to the $(Co,Zn)F_2$ (Cowley *et al.* 1975) and the success of the CPA in describing the positions and profiles of the excitations in these systems is well established.

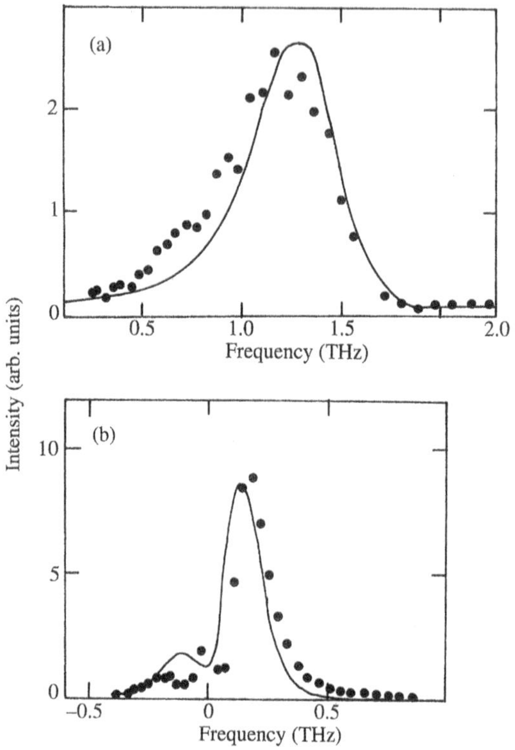

Fig. 5.7 Magnon lineshapes in $Mn_{0.78}Zn_{0.22}F_2$ compared with CPA calculations at reduced wavevectors (a) (–0.5,0,1), (b) (0,0,1). (After Buyers *et al.* 1973.)

5.4 CRITICAL SCATTERING

Critical scattering is generally quasielastic. The magnetic neutron scattering with no energy analysis is then directly related to the real part of the zero frequency dynamic susceptibility (1.42)

$$\frac{d\sigma}{d\Omega} = 2V\left(\frac{e\gamma}{\hbar c}\right)^2 k_B T \sum_{\alpha} \chi'_{\alpha\alpha}(\boldsymbol{\kappa}, 0). \tag{5.20}$$

The neutron scattering can then be used to measure the critical exponents of the susceptibility or, in the case of antiferromagnets, the staggered susceptibility. Below the critical temperature the critical scattering is contaminated by the elastic defect scattering coming from the static inhomogeneous distribution of moment. However this is very small very close to the critical temperature.

Most interest has centred on systems which can be described by the Ising model because much theoretical work has been done on this model for both pure and mixed systems. The various estimates of the critical exponents have also shown differences between the pure and the mixed systems.

The antiferromagnet FeF_2 is good example of an Ising magnet with the ratio of the exchange field to the anisotropy field acting on each spin 0.35 (Birgeneau *et al.* 1983). The two critical exponents which can be determined from neutron scattering are γ the exponent for the divergence of the staggered susceptibility and ν the exponent for the divergence of the correlation length. In addition, there is an asymmetry around the critical temperature in which amplitude of the susceptibility and the magnitude of the correlation length are different above and below the the critical temperature. The first three of these were determined for FeF_2 by Hutchings *et al.* (1972) and the exponents and the susceptibility amplitude ratio (Table 5.2) are in reasonable agreement with theoretical estimates for the pure Ising system.

The measurements for the site diluted system were done (Birgeneau *et al.* 1983) on $Fe_{0.5}Zn_{0.5}F_2$. This composition ensured that the randomness was a maximum and yet the system was well away from the critical composition for antiferromagnetism. The effects of randomness were expected to show up at reduced temperatures $t = |T/T_N - 1| \sim 10^{-2}$ so it was important to have a crystal in which the composition and therefore the critical temperature was well defined. Birgeneau *et al.* (1983) had to mask their crystal to expose a volume which had $T_N = 42.5 \pm 0.05$ K determined from monitoring the Bragg intensity. This then allowed reduced temperatures down to $t = 10^{-3}$ to be measured.

Figure 5.8 shows transverse scans above and below the Néel temperature through the (1,0,0) reciprocal lattice position. The mosaic spread of the crystal was small and accurately Gaussian. The fits in Fig. 5.8 are of a Lorentzian convoluted with the instrumental resolution. This assumes that the susceptibility has the form

$$\chi'(\boldsymbol{\kappa},0) = \frac{\chi'(\boldsymbol{\tau},0)}{1+\dfrac{\kappa^2}{\kappa_0^2}} \tag{5.21}$$

where $\chi'(\boldsymbol{\tau},0)$ is the staggered static susceptibility at the antiferromagnetic reciprocal lattice point $\boldsymbol{\tau}$ and $\boldsymbol{\kappa}_0$ is the inverse correlation length. Both of these are fitted parameters of the scans at each temperature. Figure 5.9 shows the variations of the fitted parameters as a function of reduced temperature above

Table 5.2 Exponents and susceptibility and correlation length ratios for pure and random three-dimensional Ising systems

	Pure Ising	FeF_2	Random Ising	$Fe_{1-C}Zn_CF_2$ C=0.4, 0.5	$Fe_{0.46}Zn_{0.54}F_2$
γ	1.24	1.38±0.08	1.34	1.44±0.06	1.31±0.03
ν	0.63	0.67±0.04	0.70	0.73±0.03	0.69±0.01
χ_{0+}/χ_{0-}	5.1	6.1±1.0	1.7	2.2	2.8±0.2
κ_{0+}/κ_{0-}			0.83		0.69±0.02

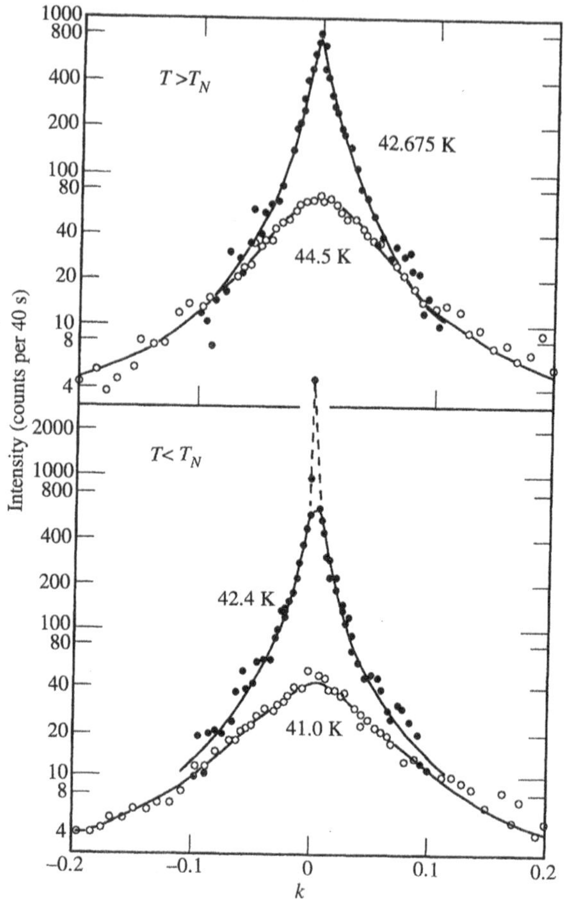

Fig. 5.8 Transverse scans through the (1,0,0) Bragg peak position in $Fe_{0.5}Zn_{0.5}F_2$ above and below the Néel temperature of 42.50 K. The full lines are Lorentzian fits. (After Birgeneau *et al.* 1983.)

and below T_N which can be fitted to give the scaling exponents γ and ν. These exponents, within error, were found to be the same both sides of T_N and are shown in Table 5.2 along with the theoretical estimates. What is also obvious from Fig. 5.9 is that there is considerable difference between the magnitudes of the parameters $\boldsymbol{\kappa}_0$ and $\chi'(G)$ in the temperature ranges above and below T_N. The ratios between the two temperature ranges are also given in Table 5.2 where it is seen that the ratio of susceptibility amplitudes is quite different to that found for pure FeF_2 and also quite different to the theoretical result for the pure Ising system, but similar to the theoretical estimate for the random exchange Ising model. Also shown in the last column of Table 5.2 are further experimental determinations (Belanger *et al.* 1986) of the scaling exponents and the ratios of

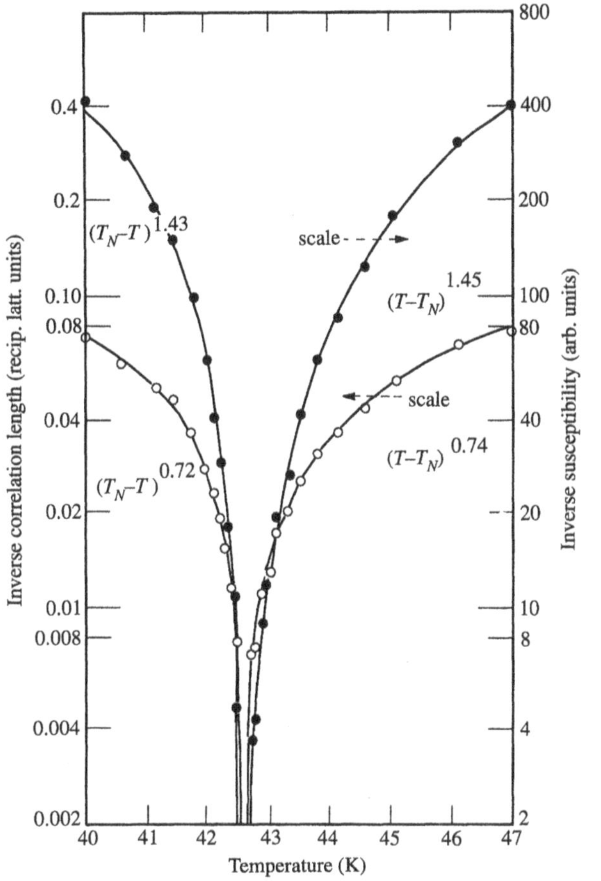

Fig. 5.9 Inverse correlation length and inverse staggered susceptibility in $Fe_{0.5}Zn_{0.5}F_2$ above and below T_N. The solid lines are the results of the fit of single power laws. (After Birgeneau *et al.* 1983.)

susceptibility and inverse correlation length in a more uniform crystal of (Fe,Zn)F_2.

Different critical behaviour again is evident when the diluted Ising systems are subject to a magnetic field. Briefly, a three-dimensional diluted Ising antiferromagnet in a field is at its lower critical dimension and its state depends on its previous magnetic history (Birgeneau *et al.* 1985). The effect of this on the critical scattering is dramatically evident in results from the more uniform $Fe_{0.46}Zn_{0.54}F_2$ crystal mentioned above (Belanger *et al.* 1987). Fig. 5.10 (inset) shows the effect of warming through the critical region at two different rates after cooling to below the critical temperature and applying the field. The slower the traverse the more developed the critical scattering. Figure 5.10 also shows that the development of the critical scattering is logarithmic in time and the variation is shown for various temperatures and fields after the zero field cooling procedure.

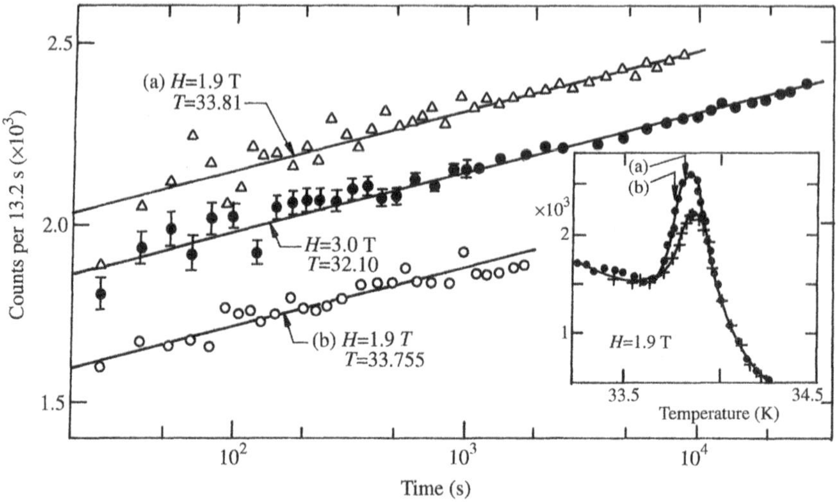

Fig. 5.10 The time variation of the intensity at (1,0,0) in $Fe_{0.46}Zn_{0.54}F_2$ at several different fields and temperatures. Inset shows the variation of intensity with temperature for two different scan rates. (After Belanger *et al.* 1987.)

This is a good example of a phenomenon in which the behaviour in the random system is quite different from that in the pure system.

This chapter has touched on the rich variety of phenomena associated with magnetic excitations in disordered alloys with magnetic long range order. Although the experiments covered are a small minority of those in the literature it is clear that many of the explanations need concepts which are universal in theoretical physics and are near the boundaries of the discipline.

References

Abrahams, S.C., Guttman, L., and Kasper, J.S. (1962). *Phys. Rev.* **127**, 2052.

Belanger, D.P., King, A.R., and Jaccarino, V. (1986). *Phys. Rev. B* **34**, 452.

Belanger, D.P., Jaccarino, V., King, A.R., and Nicklow, R.M. (1987). *Phys. Rev. Lett.* **59**, 930.

Birgeneau, R.J., Cowley, R.A., Shirane, G., Yoshizawa, H., Belanger, D.P., King, A.R. *et al.* (1983). *Phys. Rev.* B **27**, 6747.

Birgeneau, R.J., Cowley, R.A., Shirane, G., and Yoshizawa, H. (1985). *Phys. Rev. Lett.* **54**, 2147.

Buyers, W.J.L., Holden, T.M., Svensson, E.C., and Cowley, R.A. (1971). *Phys. Rev. Lett.* **27**, 1442.

Buyers, W.J.L., Pepper, D.E., and Elliott, R.J. (1972). *J. Phys. C* **5**, 2611.

Buyers, W.J.L., Pepper, D.E., and Elliott, R.J. (1973). *J. Phys. C* **6**, 1933.

Cowley, R.A. and Buyers, W.J.L. (1972). *Rev. Mod. Phys* **44**. 406.

Cowley, R.A., Dietrich, O.W., and Jones, D.A. (1975). *J. Phys. C* **8**, 3023.

Hatherley, M., Hirakawa, K., Lowde, R.D., Mallett, J.F., Stringfellow, M.W., and Torrie, B.H. (1964). *Proc. Phys.* Soc. **84**, 55.
Hennion, M., Hennion, B., and Kajzar, F. (1977). *Solid State Comm.* **21**, 231.
Hutchings, M.T., Schulhof, M.P., and Guggenheim, H.J. (1972). *Phys. Rev. B* **5**, 154.
Lowde, R.D., Shimizu, M., Stringfellow, M.W., and Torrie, B.H. (1965). *Phys. Rev. Lett.* **14**, 698.

6

Spin glasses

Spin glasses are spatially magnetically disordered materials which are frozen at least on the time scale of ordinary laboratory measurements. This means that the magnetic fluctuations of these materials in the frozen state are definitely much slower than can be detected with neutrons. However, the static configuration of magnetic moments gives rise to magnetic diffuse scattering of neutrons which can be isolated by the same polarization analysis techniques outlined at the beginning of Chapter 4 on antiferromagnets. Even above the freezing temperature, measured at laboratory times by, say, the acquisition of susceptibility data as a function of temperature, the magnetic fluctuations are so slow that quasi-elastic approximations for the neutron scattering can be used and a picture of the magnetic moment correlations can be obtained which if not static on laboratory time scales are static on neutron scattering time scales.

The freezing process for spin glasses is of considerable interest and it turns out that neutron scattering can shed some light on this subject, firstly by providing a measurement of the effective freezing temperature at times much shorter than normal laboratory times, and secondly by following the spectrum of magnetic fluctuations as a function of temperature. This latter activity has been stretched to longer times by neutron spin echo spectroscopy which naturally incorporates polarization analysis. Both the static and the dynamic measurements will be described in this chapter, but firstly we need to establish the relationship between magnetic fluctuations and the magnetic susceptibility.

6.1 SPIN CORRELATIONS AND SUSCEPTIBILITY

Much of the neutron scattering literature on spin glasses has been concerned with the spatial correlations between moments. We are concerned here with atomically disordered paramagnets with energy levels which are very closely spaced. We restrict ourselves to materials which do not have a preferred magnetic direction so that we can omit the second term in the expression for the magnetic cross section without analysis [eqns (1.31) and (1.32)]. Then the magnetic cross section can be written

$$\frac{\mathrm{d}\sigma}{\mathrm{d}\Omega\mathrm{d}E}=\left(\frac{e\gamma}{\hbar c}\right)^2\frac{k'}{k}\sum_{\alpha}\left(1-\tilde{\kappa}_{\alpha}^2\right)\sum_{q}\sum_{q'}P_q\left|\left\langle q'\left|\hat{\boldsymbol{M}}_{\alpha}(\boldsymbol{\kappa})\right|q\right\rangle\right|^2\delta\left(\hbar\omega+E_{q'}-E_q\right) \tag{6.1}$$

where $\tilde{\kappa}_\alpha$ is the component of the unit scattering vector along the α direction. We shall find this nomenclature more convenient for this section. If the moments are localized spins at fixed locations, the $\hat{M}_\alpha(\boldsymbol{\kappa})$ can be written

$$\hat{M}_\alpha(\boldsymbol{\kappa}) = g\mu_B \sum_{R} \hat{S}_\alpha(\boldsymbol{R})\mathrm{e}^{\mathrm{i}\boldsymbol{\kappa}\cdot\boldsymbol{R}} \tag{6.2}$$

where the $S(\boldsymbol{R})$ are effective spins with the local moments $g\mu_B S$. For scattering with very little energy change the sum over q' can be performed by closure and

$$\frac{\mathrm{d}\sigma}{\mathrm{d}\Omega} = \left(\frac{e^2\gamma}{2m_e c^2}\right)^2 g^2 \sum_{i,j}\sum_{\alpha}\left(1-\tilde{\kappa}_\alpha^2\right)\overline{\left\langle \hat{S}_\alpha(\boldsymbol{R}_j)\hat{S}_\alpha(\boldsymbol{R}_i)\right\rangle}\, \mathrm{e}^{\mathrm{i}\boldsymbol{\kappa}\cdot(\boldsymbol{R}_i-\boldsymbol{R}_j)} \tag{6.3}$$

where the brackets signify expectation value and the bar is the thermal average over the states q. In the case of a magnetic alloy with one magnetic species the distribution of the magnetic species must be taken into account. To do this the occupation operator $p(\boldsymbol{R})$ may be used in a similar fashion to that for ferromagnetic and antiferromagnetic alloys.

$$\frac{\mathrm{d}\sigma}{\mathrm{d}\Omega} = \left(\frac{e^2\gamma}{2m_e c^2}\right)^2 g^2 \sum_{i,j}\sum_{\alpha}\left(1-\tilde{\kappa}_\alpha^2\right)\overline{p(\boldsymbol{R}_j)p(\boldsymbol{R}_i)\left\langle \hat{S}_\alpha(\boldsymbol{R}_j)\hat{S}_\alpha(\boldsymbol{R}_i)\right\rangle}\, \mathrm{e}^{\mathrm{i}\boldsymbol{\kappa}\cdot(\boldsymbol{R}_i-\boldsymbol{R}_j)} \tag{6.4}$$

where the bar represents an average over the distribution of magnetic species as well as the thermal average. The configurational average may be done separately if there is little correlation between the distribution of the magnetic species and the thermally averaged correlation between spins.

$$\frac{\mathrm{d}\sigma}{\mathrm{d}\Omega} = \left(\frac{e^2\gamma}{2m_e c^2}\right)^2 g^2 \sum_{i,j}\sum_{\alpha}\left(1-\tilde{\kappa}_\alpha^2\right)\left[C^2 + C(1-C)\alpha(\boldsymbol{R}_i-\boldsymbol{R}_j)\right]\overline{\left\langle S_\alpha(\boldsymbol{R}_j)S_\alpha(\boldsymbol{R}_i)\right\rangle}\, \mathrm{e}^{\mathrm{i}\boldsymbol{\kappa}\cdot(\boldsymbol{R}_i-\boldsymbol{R}_j)} \tag{6.5}$$

with $\alpha(\boldsymbol{R}-\boldsymbol{Q})$ the Cowley short range order parameter. This form was first used by Wells and Smith (1971) and can be more conveniently written by taking the expectation values of the spin operators

$$\frac{\mathrm{d}\sigma}{\mathrm{d}\Omega} = \frac{2}{3}\left(\frac{e^2\gamma}{2mc^2}\right)^2 g^2 C\left\{ S(S+1) + S^2 \sum_{R_i, R_j \neq R_i}\left[C + (1-C)\alpha(\boldsymbol{R}_i-\boldsymbol{R}_j)\right]\gamma(\boldsymbol{R}_i-\boldsymbol{R}_j)\mathrm{e}^{\mathrm{i}\boldsymbol{\kappa}\cdot(\boldsymbol{R}_i-\boldsymbol{R}_j)}\right\} \tag{6.6}$$

for a cubic system, with S now the spin quantum number, and γ the spin pair correlation function defined so that $\gamma = 1$ for perfect correlation.

Another approach which yields further insight, is to start from the the formulation of the quasielastic cross section in terms of the wavevector dependent susceptibility (1.42)

$$\frac{\mathrm{d}\sigma}{\mathrm{d}\Omega} = 2V\left(\frac{e\gamma}{\hbar c}\right)^2 k_B T \ \chi'(\boldsymbol{\kappa},0) \tag{6.7}$$

where $\chi'(\boldsymbol{\kappa},0)$ is the $\boldsymbol{\kappa}$ dependent real part of the dynamic susceptibility at zero frequency. For a noninteracting system with a concentration C of magnetic sites the susceptibility is simply

$$\chi'(\boldsymbol{\kappa},0) = g^2\mu_B^2\, f^2(\boldsymbol{\kappa})C\frac{S(S+1)}{3k_BT} = f^2(\boldsymbol{\kappa})C\chi_0. \tag{6.8}$$

In the presence of interactions an approximation to the susceptibility, accurate at high temperatures, can be obtained using a local molecular field model. Within this model the moment at any site can be written

$$\mu(\boldsymbol{R}) = p(\boldsymbol{R})\chi_0\left\{h_a(\boldsymbol{R}) + \frac{1}{g^2\mu_B^2}\sum_{\boldsymbol{R}'\neq\boldsymbol{R}} j(\boldsymbol{R}-\boldsymbol{R}')\mu(\boldsymbol{R}')\right\} \tag{6.9}$$

where $h_a(\boldsymbol{R})$ is the applied field at $\boldsymbol{R}$, and where the second term provides a local molecular field dependent on the local magnetic environment through an exchange interaction $j(\boldsymbol{R} - \boldsymbol{R}')$. Taking a configurational average of $\mu(\boldsymbol{R})$ over the distribution of magnetic atoms gives

$$\overline{\mu(\boldsymbol{R})} = \overline{p(\boldsymbol{R})}\ \chi_0\overline{h(\boldsymbol{R})} + \frac{\chi_0}{g^2\mu_B^2}\sum_{\boldsymbol{R}'\neq\boldsymbol{R}} j(\boldsymbol{R}-\boldsymbol{R}')\ \overline{p(\boldsymbol{R})p(\boldsymbol{R}')}\ \frac{\overline{\mu(\boldsymbol{R}')}}{C} \tag{6.10}$$

in which it has been assumed that there is no correlation between the distribution of magnetic atoms and the applied field, and that part of the correlation between the presence of an atom at $\boldsymbol{R}$ and the moment at $\boldsymbol{R}'$ due to positional correlations between magnetic atoms has been taken into account in the average over the product $p(\boldsymbol{R})\mu(\boldsymbol{R}')$. Fourier transforming and taking the averages, we have

$$\boldsymbol{M}(\boldsymbol{\kappa}) = C\chi_0\boldsymbol{H}(\boldsymbol{\kappa}) + \frac{\chi_0}{g^2\mu_B^2}\boldsymbol{M}(\boldsymbol{\kappa})\sum_{\boldsymbol{R}}\left[C + (1-C)\alpha\,(\boldsymbol{R}-\boldsymbol{R}')\right]j(\boldsymbol{R}-\boldsymbol{R}')\mathrm{e}^{\mathrm{i}\boldsymbol{\kappa}\cdot(\boldsymbol{R}-\boldsymbol{R}')} \tag{6.11}$$

where α is the Cowley short range order parameter. The wavevector dependent susceptibility is then

$$\chi(\boldsymbol{\kappa},0) = \frac{\boldsymbol{M}(\boldsymbol{\kappa})}{\boldsymbol{H}(\boldsymbol{\kappa})} = \frac{Cf^2(\boldsymbol{\kappa})\chi_0}{1 - \frac{\chi_0}{g^2\mu_B^2}\left[CJ(\boldsymbol{\kappa}) + (1-C)L(\boldsymbol{\kappa})\right]} \tag{6.12}$$

where the square of the magnetic form factor, $f^2(\boldsymbol{\kappa})$, has been inserted because of the distributed nature of the moment at each site, and $L(\boldsymbol{\kappa})$ is the convolution of $J(\boldsymbol{\kappa})$ with the Fourier transform of the Cowley parameters. This form was first used by Ahmed and Hicks (1974).

6.2 THE SMALL κ SUSCEPTIBILITY OF DILUTE Cu–Mn SPIN GLASSES

The strange magnetic properties of Cu–Mn alloys were the subject of study from the late 1950s. The maximum in the susceptibility at low temperatures was observed by Owen *et al.* (1957) and history dependent magnetic effects including displaced hysteresis after cooling in a magnetic field were documented by Kouvel (1961). Early neutron diffraction investigations (Bacon *et al.* 1957; Meneghetti and Sidhu 1957) of low manganese concentration alloys observed no magnetic Bragg peaks. Even an exhaustive search of reciprocal space looking for a spin density wave using single crystals failed to find sharp features in the scattering (Arrott 1965). In a series of powder and single crystal diffraction measurements Sato and Werner and their associates identified significant diffuse scattering, especially around (1,1/2,0) positions in a well-annealed 25 at% manganese alloy which by temperature dependence they confirmed contained a significant magnetic contribution (Sato *et al.* 1974). At the same time a separation of the magnetic component was attempted in single crystals by Wells and Smith (1971) using the form factor variation of the magnetic cross section.

The first direct measurement of the diffuse magnetic cross section was by Ahmed and Hicks (1974) who used polarization analysis without energy analysis to separate out the magnetic scattering from a series of polycrystalline alloys dilute in manganese as a function of temperature. Figure 6.1 shows the scattering vector dependence of both the magnetic and nuclear diffuse cross sections. The nuclear defect cross sections are independent of temperature and for the 2 and 5 at% manganese alloys indicate that the alloys are closely random because of the lack of any $\boldsymbol{\kappa}$ dependence. For the 10 at% alloy the nuclear scattering is $\boldsymbol{\kappa}$ dependent and over the limited scattering vector range can be adequately fitted by a Cowley short range order parameter $\alpha_1 = -0.035$.

The magnetic scattering is more interesting. It is temperature dependent and also tends to peak towards small scattering vectors. Ahmed and Hicks (1974) used the relationship (6.7) to analyse their results. $\chi(0,0)$ is the bulk d.c. magnetic susceptibility measured in a uniform field, and thus a comparison can be made between bulk measurements and the extrapolation to $\boldsymbol{\kappa} = 0$ from the magnetic diffuse scattering. This comparison is shown in Fig. 6.1, where the points at $\boldsymbol{\kappa} = 0$ come from bulk susceptibility measurements. It is clear that for the higher temperature results the comparison is good, but for lower temperatures the bulk susceptibility measurements lie below any reasonable extrapolation of the neutron scattering data. This can only mean that at low temperatures the bulk measurements are not accessing some of the susceptibility or that some of the spatial fluctuations in magnetization seen in the neutron scattering experiment are not dynamic and therefore not connected with a susceptibility. Either way the low temperature state, if it has no long range order, is frozen at least on the time scale of a laboratory 'd.c.' susceptibility measurement. This underscores the validity of the quasistatic assumption made in eqn (6.7).

One of the significant features of $\chi(\boldsymbol{\kappa})$ determined by Ahmed and Hicks (1974) is the absence of any abrupt change in spatial spin correlation at the tem-

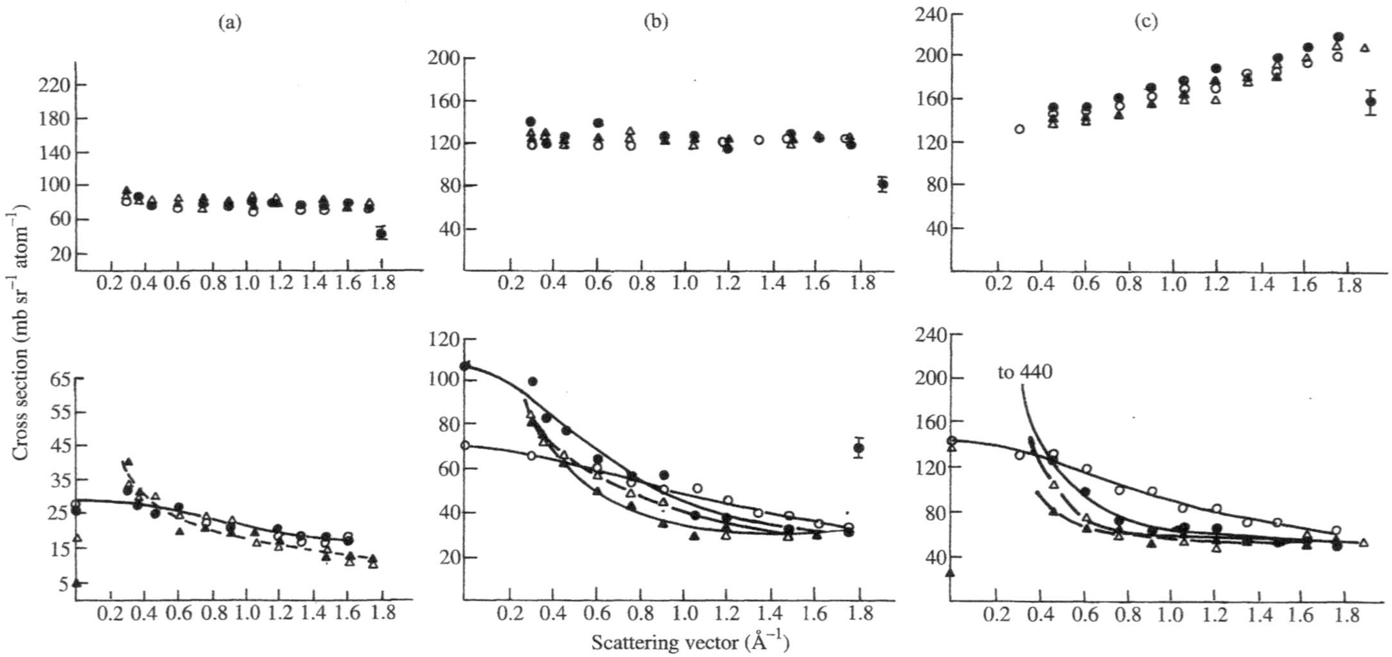

Fig. 6.1 Neutron scattering cross section for three Cu–Mn spin glasses. The upper plots show the nuclear scattering and the lower plots show the temperature and wavevector dependence of the magnetic scattering (a) 2 at% Cu–Mn (○) 40 K, (●) 30 K, (Δ) 15 K, (▲) 4.2 K; (b) 5 at% Cu–Mn (○) 293 K, (●) 35 K, (Δ) 15 K, (▲) 4.2 K; (c) 10 at% Cu–Mn (○) 293 K, (●) 90 K, (Δ) 19 K, (▲) 4.2 K. Also shown are the $\boldsymbol{\kappa} = 0$ points corresponding to the bulk susceptibility at the various temperatures on the same samples. The glass temperatures for the three alloys are approximately 14, 29, and 43 K. (After Ahmed and Hicks 1974.)

perature of the maximum in the susceptibility. In a normal magnetic phase transformation the range of the correlation diverges as the critical temperature is approached. In this transformation virtually nothing changes spatially but there is apparently a large change in the response time of the system.

It is interesting that the magnetic scattering observed by Ahmed and Hicks (1974) reveals predominantly ferromagnetic coupling. This contrasts with the antiferromagnetic long range order which is eventually established at much higher manganese concentrations. Ahmed and Hicks (1974) attempted to extract exchange interactions from their data by applying eqn (6.12) to the observed $\chi(\boldsymbol{\kappa})$ in the paramagnetic phase at high temperatures. The 5 at% manganese sample was found to be closely random so that some estimate could be made of the spherical average of $J(\boldsymbol{\kappa})$ within the local mean field approximation. The 10 at% manganese alloy had some short range atomic order so that the part of $J(\boldsymbol{\kappa})$ convoluted with the atomic distribution could not be obtained explicitly. Rough as this procedure is it seems to indicate, from the range of $J(\boldsymbol{\kappa})$ that the ferromagnetic exchange is largely first neighbour. Figure 6.2 shows the variation of the exchange interaction with $\boldsymbol{\kappa}$. The value of $\boldsymbol{\kappa}$ for which $J(\boldsymbol{\kappa})$ crosses the axis is that expected for a predominantly first neighbour interaction which would have to be ferromagnetic. This does not preclude some more complicated combination of interactions which combine to give the observed $\boldsymbol{\kappa}$ variation and of course relies on the applicability of the mean field approximation. Davis and Hicks (1979) have measured the scattering from a 4.7 at% manganese alloy single crystal at 4.2 K in the spin glass phase, and in spite of a restricted $\boldsymbol{\kappa}$ range the variation of the magnetic cross section with $\boldsymbol{\kappa}$ in a number of directions also supports a ferromagnetic first neighbour interaction. Susceptibility measurements (Morgownik and Mydosh 1981) also endorse the theory that the interaction is largely first neighbour and ferromagnetic with an estimate for the strength of the first neighbour interaction within a factor two of that obtained by Ahmed and Hicks (1974). In

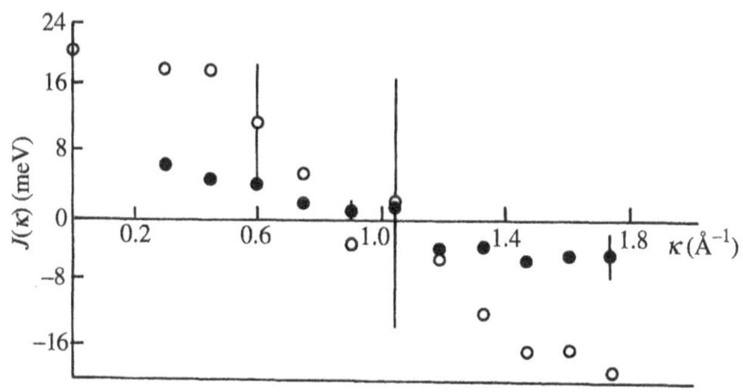

Fig. 6.2 The wavevector exchange interaction $J(\boldsymbol{\kappa})$ for 5 at% Cu–Mn. (○) 293 K, (●) 35 K. The point at $\boldsymbol{\kappa}$ =0 comes from the Curie–Weiss analysis of the high temperature bulk susceptibility. (After Ahmed and Hicks 1974.)

contrast, the work of Davis *et al.* (1980) yields a strongly antiferromagnetic first neighbour interaction, but this analysis may be flawed because eqn (6.5) was used in the presence of both strong magnetic and atomic correlations. Further, only relative spin correlations were reported and the absolute first and second neighbour spin correlations obtained by refitting the data appear to be better than perfect. The Au–Fe system has similar contradictions. There the spin glass region gives way to long range ferromagnetism near the percolation concentration for first neighbour interactions (Murani 1980), yet it is well established that the face centred cubic form of iron is antiferromagnetic (Tsunoda 1989).

6.3 LARGER κ FEATURES IN Cu–Mn ALLOYS

At higher scattering vectors there are further features which complicate the interpretation of the magnetic scattering. These features, found by Werner and Cable (1981), appear at positions in reciprocal space shown in Fig. 6.3. The open circles mark broad features in the nuclear defect scattering, and crosses mark the positions for magnetic features. The magnetic and nuclear diffuse cross sections were separated using polarization analysis. The variation of differential cross section is shown in Fig. 6.4 in scans through the (1,1/2,0) type positions. There is a large increase in the scattering as the manganese concentration is increased. The magnetic cross section is temperature dependent and this is shown for the 15 at% manganese alloy in Fig. 6.4. It should be emphasized that the magnetic scattering is never sharp and at most refers to correlations which have a range of 30–40 Å, even for 25 at% manganese. As the temperature is increased the features in the magnetic scattering gradually decrease in amplitude and broaden retaining their basic shape to many times the glass temperature. It is clear from

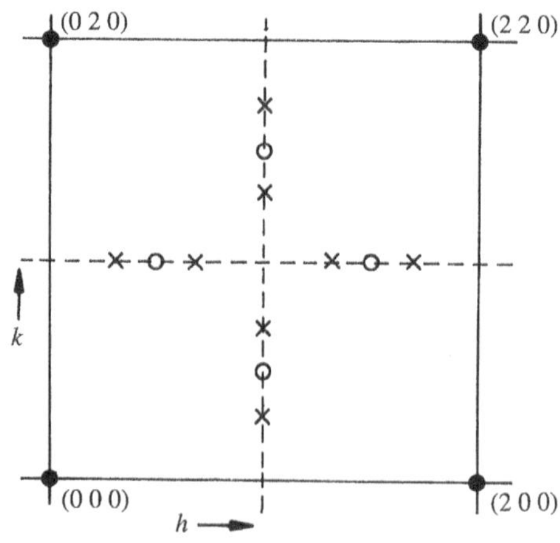

Fig. 6.3 Positions in reciprocal space where broad magnetic features (○) and sharper magnetic features (×) are observed. (After Werner and Cable 1981.)

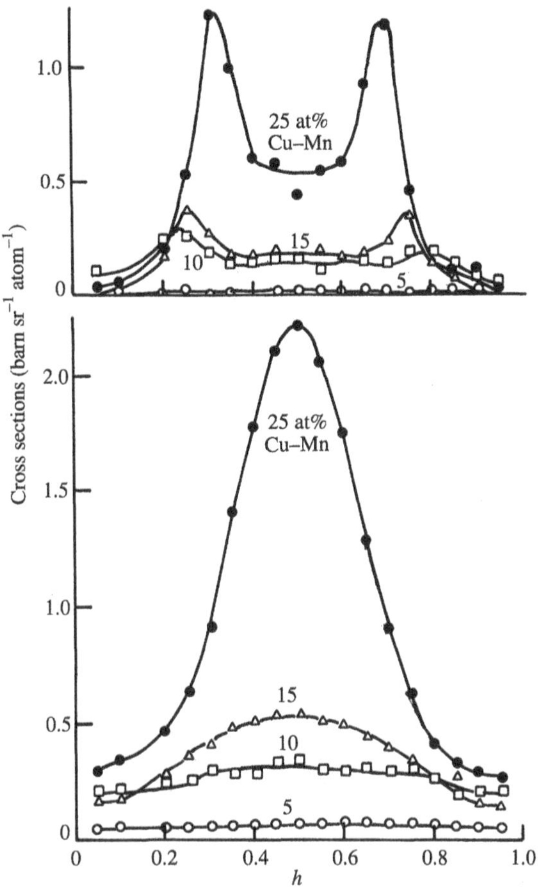

Fig. 6.4 Magnetic (upper) and nuclear (lower) neutron cross sections for Cu–Mn alloys along the line [*h*,1,0] in **Fig.** 6.3. (After Werner and Cable 1981.)

this work that there are magnetic periodicities which coexist with the tendency for ferromagnetic alignment. It is most probable that the wavevector of these periodicities is related to the dimensions of the Fermi surface which changes with alloy concentration (Harders and Wells 1983).

In addition to the scattering around the (1,1/2,0) positions, the scattering at small angles persists. For instance the work of Gray *et al.* (1982) shows strong small angle scattering for a 16.7 at% manganese alloy with long range magnetic clusters with an average radius of gyration of 26 Å containing about 170 manganese atoms. This suggests that clusters containing the antiferromagnetic and the ferromagnetic correlations may be one and the same, but Cable *et al.* (1984) have shown that the width of the sharper of the magnetic features near (1,1/2,0) is virtually independent of concentration so that at 5 per cent

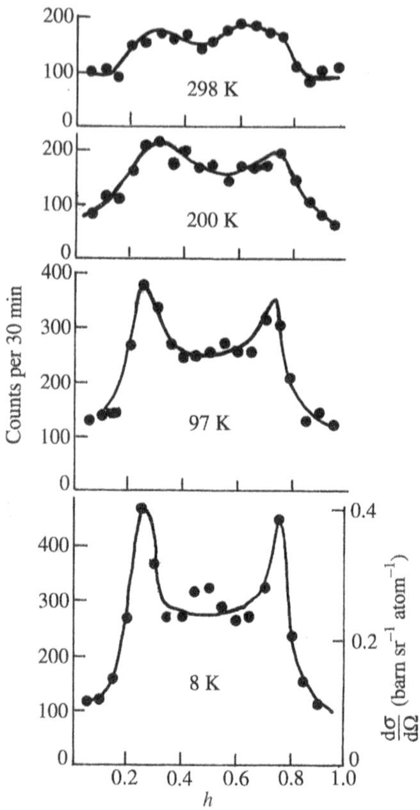

Fig. 6.5 The temperature variation of the features in the magnetic neutron scattering for 15 at% Cu–Mn. (After Werner and Cable 1981.)

manganese it is sharper than any feature observed at small **κ**. Even so, Cable *et al.* (1984) have attempted to fit the broader of the two magnetic features as ferromagnetic defect scattering assuming long range *ferromagnetic* correlations, and the broad rise in the magnetic scattering at small **κ** as due to the reflection of long range *antiferromagnetic* correlations in the magnetic defect scattering of a psuedo-antiferromagnet with atomic short range order. Both the small **κ** increase in magnetic cross section and the diffuse peak from the (1,1/2,0) region is seen by Harders *et al.* (1983) using polarization analysis to separate out the magnetic scattering in alloys containing 10 and 20 at% manganese, but again attempts to fit both peaks are not very successful.

6.4 Au–Fe SPIN GLASSES

The Au–Fe spin glass is characterized by long range ferromagnetic correlations and at small wavevectors, in relatively concentrated alloys of 10 and 13 at%

iron, a cusp (Fig. 6.6) can be observed in the variation of the scattering with temperature (Murani 1976). In fact the temperature of the cusp in the scattering changes with scattering vector, and this was interpreted by Murani (1976) to mean that magnetic clusters on different scales have different freezing tempera-

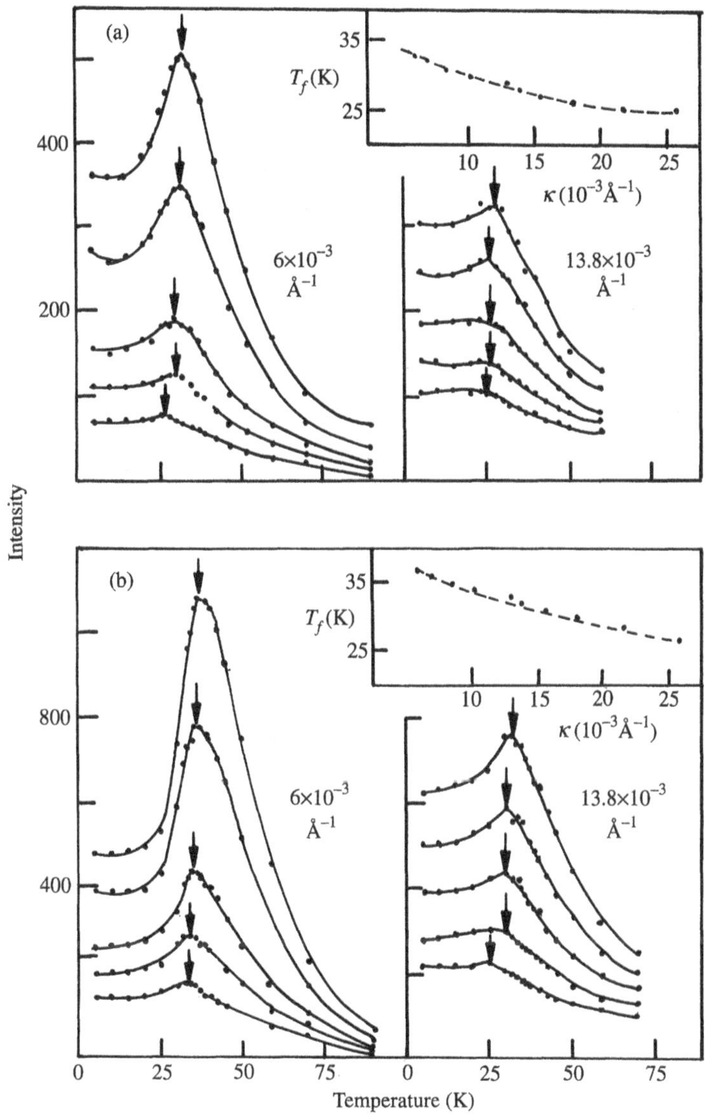

Fig. 6.6 The forward scattering intensity as a function of temperature for a series of **κ** values for (a) Au–Fe 10 at% (b) Au–Fe 13 at%. The peaks in the scattering are plotted against **κ** in the insets. (After Murani 1976.)

tures. This has been disputed by Soukoulis *et al.* (1978) who argue that the data is consistent with a unique spin glass transition temperature. They break up the cross section into an elastic part, with a Brillouin-like variation with temperature, and an inelastic part, and demonstrate that with a peak at constant temperature in the inelastic part of the cross section, the peak in the total cross section can change with scattering vector. The arguments are too qualitative for firm conclusions to be drawn.

Au–Fe, like Cu–Mn, has a feature in the magnetic scattering at the (1,1/2,0) position (Cable *et al.* 1989). It is uncertain whether there is a corresponding feature in the nuclear scattering because the contrast in scattering lengths between gold and iron is quite small. Figure 6.7 shows the scattered neutron cross section at two temperatures along the [2 1 0] direction in a 15 at% iron crystal. The variation of the neutron intensity with temperature clearly labels it as magnetic. A similar variation of intensity with position in reciprocal space is observed with X-rays. Direct comparison of the X-ray data and the magnetic neutron data is complicated by the perturbation of the diffraction patterns by the lattice distortion effect inherent in an alloy with constituents of such different atomic sizes. This effect should be different for the two patterns. If at low temperatures the moments are as well correlated over large distances as the small angle results would suggest, the neutron diffuse scattering should mirror the

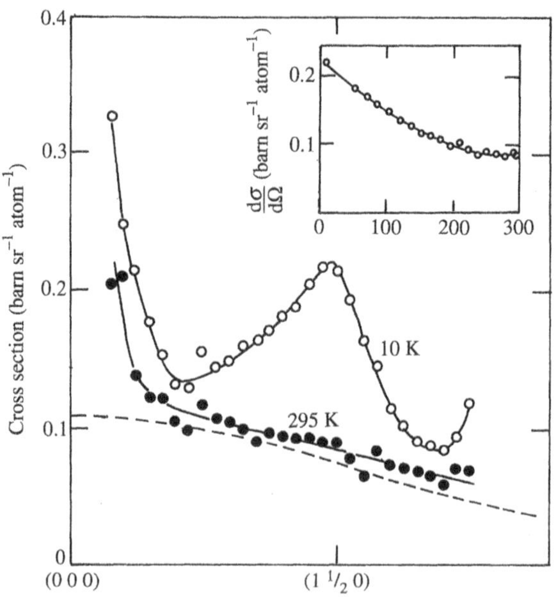

Fig. 6.7 The absolute cross section along [2 1 0] at T=10 and 295 K for Au–Fe 15 at%. The dashed curve corresponds to free spins with S=3/2. The inset shows the temperature dependence at (1 1/2 0). (After Cable *et al.* 1987.)

distribution of iron atoms. However the effect of distortion on the diffraction patterns should be quite different, for in the X-ray case the smaller atomic size is related to the smaller scattering factor for iron, whereas in the magnetic neutron case the only magnetic scattering length is associated with the smaller iron atoms. The result is that, as for Cu–Mn there is no clear idea of the origin of the magnetic peak at (1,1/2,0).

6.5 DYNAMICS OF Cu–Mn SPIN GLASSES

Murani and Tholence (1977) analysed the energy change of neutrons scattered from an 8 at% Cu–Mn sample. In their analysis they only counted neutrons outside their energy resolution as contributing to $\chi(\boldsymbol{\kappa})$, and effectively treated the rest of the scattered intensity as *strictly* elastic. They compared $\chi(\boldsymbol{\kappa})$, obtained in this way for several small values of $\boldsymbol{\kappa}$, with the low field a.c. susceptibility measured on the same sample at 16 Hz, and Fig. 6.8 shows the two sets of results. The main feature is the cusp in $\chi(\boldsymbol{\kappa})$ which is evident at a temperature

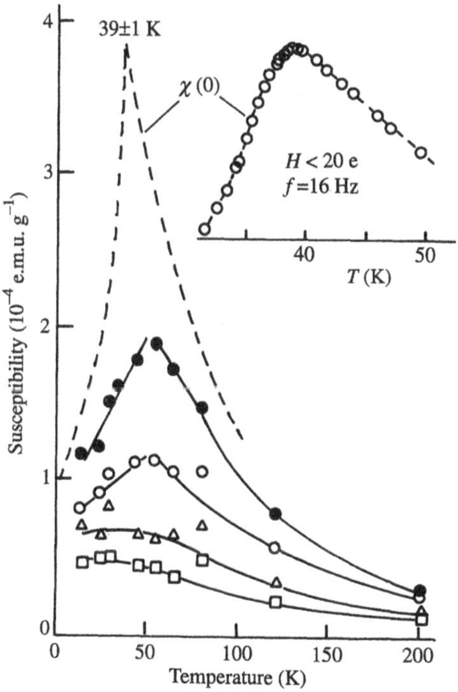

Fig. 6.8 The variation with temperature of the susceptibility calculated from that part of the neutron scattering lying outside the energy resolution (a time constant of about 10^{-11} s) for 8 at% Cu–Mn. (●) $\boldsymbol{\kappa}$ = 0.08 Å^{-1}, (○) $\boldsymbol{\kappa}$ = 0.2 Å^{-1}, (Δ) $\boldsymbol{\kappa}$ = 0.4 Å^{-1}, (□) $\boldsymbol{\kappa}$ = 0.6 Å^{-1}. The dashed curve and inset show the low frequency susceptibility. (After Murani and Tholence 1974.)

very much higher in the neutron results than in the a.c. susceptibility. The invariance of the temperature of the cusp in $\chi(\boldsymbol{\kappa})$ with $\boldsymbol{\kappa}$ suggests that the unusually high temperature would probably be a property of $\chi(0)$ derived in the same way.

The most probable explanation for the results of Murani and Tholence (1977) is that the missing part of the susceptibility is included in that part of the scattering not resolvable from the strictly elastic scattering (or there may be no strictly elastic scattering). The energy resolution of this experiment is equivalent to being able to excite those parts of the susceptibility with relaxation times less than 10^{-11} s, with longer relaxation times appearing frozen. The temperature at which a large part of the system appears frozen is higher than the 39 K for an experiment on a time scale of seconds to minutes. Alloul *et al.* (1988) have also measured the inelasticity of scattering from a more concentrated 20 at% manganese sample, but identify the scattering within their similar energy resolution with the strictly elastic scattering expected from the spin glass phase in the Edwards and Anderson (1975) model. This 'elastic scattering' also decreases as T^2 as would be expected for the Edwards–Anderson order parameter. Unfortunately the scattering identified as elastic certainly cannot be all elastic, as was shown in the results of Murani and Tholence (1977) and is shown in the following results taken at much higher energy resolution.

Neutron data do exist for the time scale 10^{-11} to 10^{-8} seconds (Mezei and Murani 1979). The neutron spin echo technique directly measures the decay of the spin correlation function in real time. The results for 5 at% manganese are shown in Fig. 6.9 for $\boldsymbol{\kappa} = 0.093$ Å^{-1}. Over most of the temperature and time range the relaxation is nowhere near exponential and is better described as being proportional to ln t. At low temperatures, even after 10^{-8} s, there is considerable remanence which shows every indication of persisting over many longer decades of time. Mezei and Murani (1979) also looked at the variation of spatial correlation with temperature in the 5 at% manganese sample. Figure 6.10 shows the variation of magnetic neutron scattering cross section, separated using polarization analysis, with temperature for two scattering vectors. At low

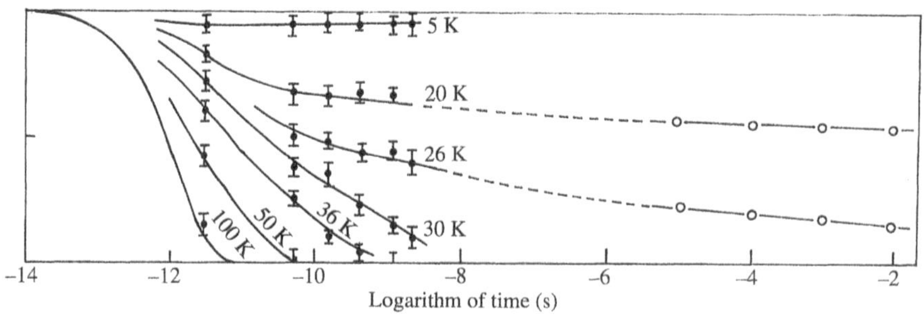

Fig. 6.9 The time dependent spin correlation function for Cu–Mn 5 at%, $e^{-\Gamma t}$ ($\Gamma = 0.5$ meV), for $\langle\boldsymbol{\kappa}\rangle = 0.093$ Å^{-1}. The thick full line corresponds to a simple exponential decay. (After Mezei and Murani 1979.)

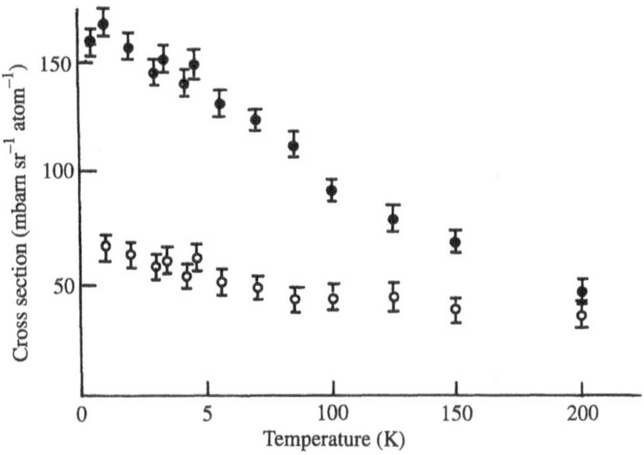

Fig. 6.10 The variation of the magnetic cross section with temperature for Cu–Mn 5 at% two scattering vectors, (●) $\langle\kappa\rangle = 0.093$ Å^{-1}, (○) $\langle\kappa\rangle = 0.37$ Å^{-1}. (After Mezei and Murani 1979.)

temperatures a ferromagnetic correlation develops as was observed by Ahmed and Hicks (1974). But more importantly, at no temperature is there any indication of a sudden change in correlation length as would be expected for a normal phase transition.

6.6 OTHER SPIN GLASSES

Europium strontium sulphide provides a well researched example of a nonmetallic spin glass system. Strontium sulphide is diamagnetic, but the addition of europium on strontium sites provides a localized moment so that for low europium concentrations the mixture is a spin glass (Maletta and Felsch 1978). Cu–Mn is thought to be a spin glass because the long range RKKY interaction (Ruderman and Kittel 1954; Kasuya 1956; Yosida 1957) mediated by the copper conduction electrons, is an oscillating function of the separation between two manganese spins. This can produce conflicting interactions at one spin from the surrounding randomly distributed spins. In $Eu_CSr_{1-C}S$ the frustration is produced by a ferromagnetic first neighbour interaction coupled with an antiferromagnetic second neighbour interaction in the face centred cubic cation lattice (Maletta and Felsch 1979). It is thought (Kasuya 1970) that these are the only important interactions.

Maletta and Felsch (1979) took diffraction patterns of powdered $Eu_{0.4}Sr_{0.6}S$ without energy analysis, and isolated the magnetic correlation which develops at low temperatures by subtracting patterns taken below and well above the freezing temperature of 1.9 K. The result is shown in Fig. 6.11, and it can be seen that strong overall ferromagnetic correlations are evident but without magnetic long range order. Unfortunately the cross sections are not on an absolute basis so that comparisons cannot be made between them and the measured susceptibility.

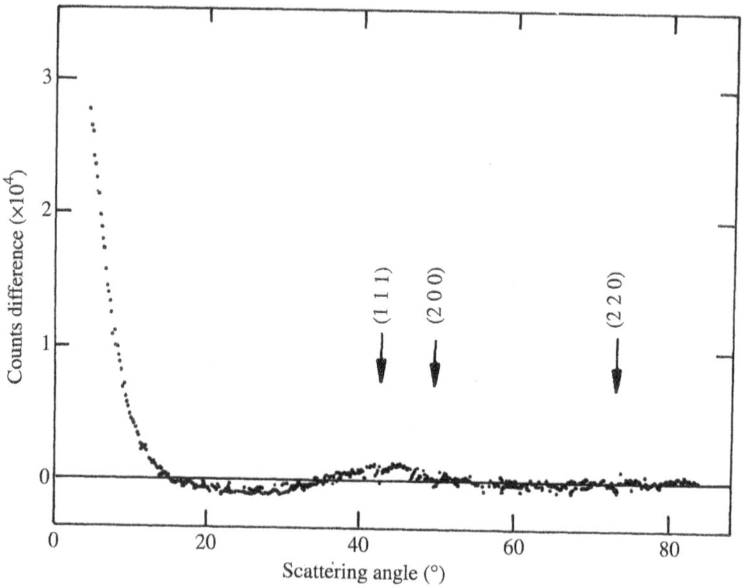

Fig. 6.11 The difference between the two diffraction patterns at 1.3. K and 10 K versus scattering angle for $Eu_{0.4}Sr_{0.6}S$. (After Maletta and Felsch 1979.)

Another extensively researched material with spin glass characteristics is $Cd_{1-C}Mn_CTe$. $Cd_{1-C}Mn_CTe$ has the zincblende structure with manganese substituting for cadmium on one of the two interpenetrating f.c.c. lattices. It exhibits strong spin correlation up to high manganese concentrations but without a transition to magnetic long range order (Giebultowicz and Holden 1988). Rather, the magnetically concentrated alloys have strong magnetic diffuse peaks at reciprocal lattice positions appropriate for type III antiferromagnetic order, but with widths which suggest that the order extends for less than about 15 cubic lattice spacings (Steigenberger *et al.* 1986). At high temperatures the correlation length increases with decreasing temperature as if approaching a normal critical point but abruptly stops at a particular temperature. At the same time the integrated intensity under the diffuse peak continues to increase with further decrease of temperature (Fig. 6.12). The diffuse peaks are also anisotropic (Steigenberger *et al.* 1986; Giebultowicz *et al.* 1988) (Fig. 6.13). At high temperatures Steigenberger *et al.* (1986) and Giebultowicz *et al.* (1988) suggest this is because the exchange interaction is anisotropic in the type III ordered structure. The strength and range of the interaction in two mutually perpendicular directions was determined by Steigenberger *et al.* (1986) using the relation (6.12). The magnetic order in these materials is so well developed that propagating magnetic excitations have been seen (Giebultowicz *et al.* 1989). The incipient antiferromagnetic order may be more complicated than type III antiferromagnetic according to a study using polarization analysis by Steigenberger and Lindley

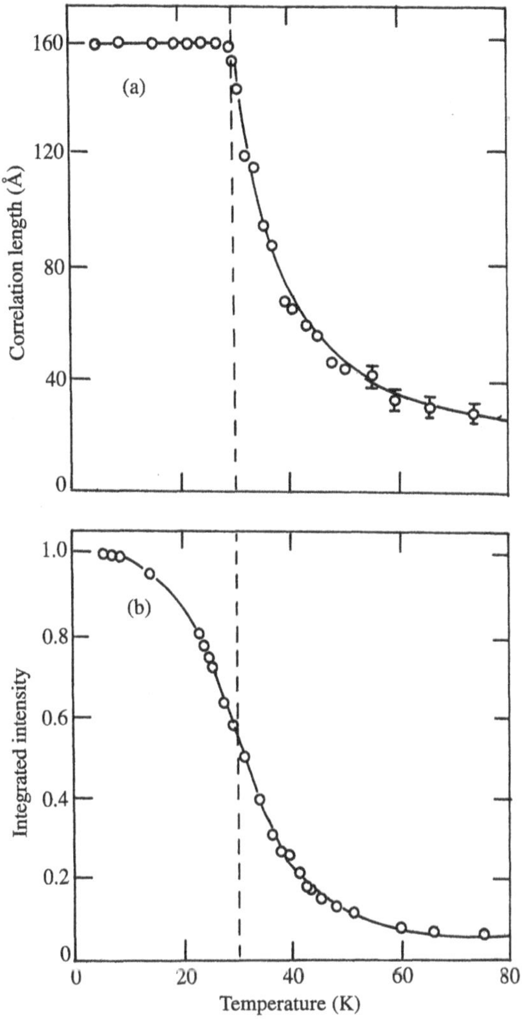

Fig. 6.12 The temperature dependence of (a) the correlation length and (b) the integrated intensity for the (1 1/2 0) magnetic peak from $Cd_{0.35}Mn_{0.65}Te$. (After Steigenberger *et al.* 1986.)

(1988) who showed that some of the magnetic diffuse peaks could produce polarization from an initially unpolarized beam. This can only happen if the incipient structure is noncollinear (Blume 1963; eqn A.14).

At the other extreme is the Ising spin glass $Fe_{1-c}Mg_{c}Cl_2$ (Wong *et al.* 1987), which has a strong uniaxial anisotropy. Figure 6.14 shows the real and imaginary parts of the measured zero field a.c. susceptibility, parallel and perpendicular to the *c*-axis of the hexagonal structure. The parallel component of both parts of the

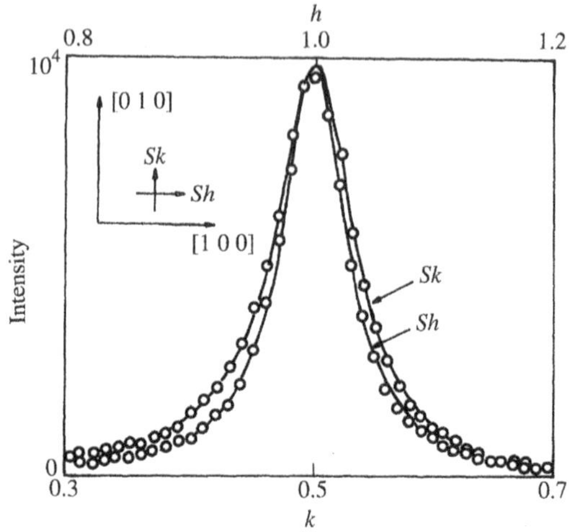

Fig. 6.13 Scans through the (1 1/2 0) position taken along [1 0 0] and [0 1 0] directions for $Cd_{0.35}Mn_{0.65}Te$ at 9.1 K. (After Steigenberger *et al.* 1986.)

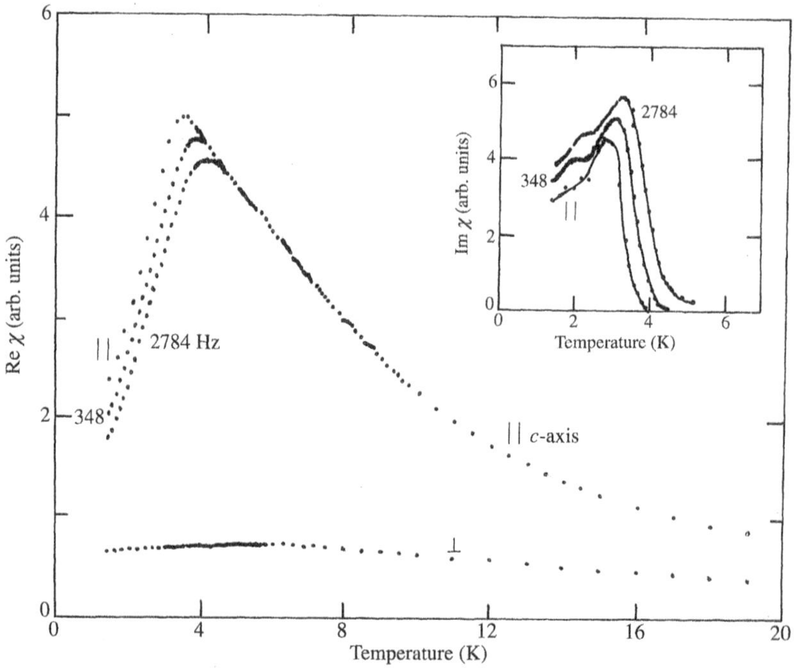

Fig. 6.14 The a.c. susceptibilities parallel and perpendicular to the easy direction in $Fe_{0.4}Mg_{0.6}Cl_2$. (After Wong *et al.* 1987.)

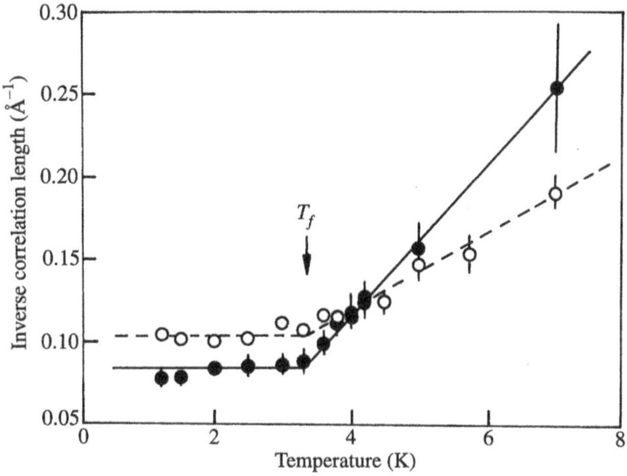

Fig. 6.15 The inverse of the spin spin correlation lengths parallel (●) and perpendicular (○) to the easy axis in $Fe_{0.4}Mg_{0.6}Cl_2$ as a function of temperature. (After Wong *et al.* 1987.)

susceptibility shows freezing at about 3.5 K depending on the frequency, whereas the perpendicular component shows no freezing. This probably means that the transverse moment fluctuations are much faster than those along the magnetic axis. Because the parallel fluctuations are slow and the scattering quasielastic it was possible to measure the parallel spin spin correlation both along and perpendicular to the easy direction. The diffuse peaks, at reciprocal lattice positions where Bragg peaks are expected for the $FeCl_2$ antiferromagnetic ordering are Lorentzian as would be predicted from eqn (6.12) with $J(\boldsymbol{\kappa})$ an even function of $\boldsymbol{\kappa}$ around these reciprocal lattice positions. Figure 6.15 shows the variation with temperature of the two correlation lengths. The variation is similar to that already seen for $Cd_{1-C}Mn_CTe$ (Fig. 6.12) and the freezing temperature can be easily seen.

6.7 RE-ENTRANT SPIN GLASSES

The first re-entrant spin glass phase to be investigated was that in Au–Fe alloys. The phase diagram (Fig. 6.16) was established by bulk magnetic measurements (Coles *et al.* 1978). For iron concentrations of more than 15.5 at% there is a temperature below the Curie temperature at which the magnetic response falls and which is interpreted as a spin glass freezing temperature. The immediate question is: does the ferromagnetic long range order survive below this temperature? It is not clear whether true long range order does survive. What is certain is that there are changes in some correlation lengths below the glass temperature. For instance the small angle scattering results of Murani (1980) show that for a 17 at% iron alloy the correlation lengths become smaller but still remain long, at

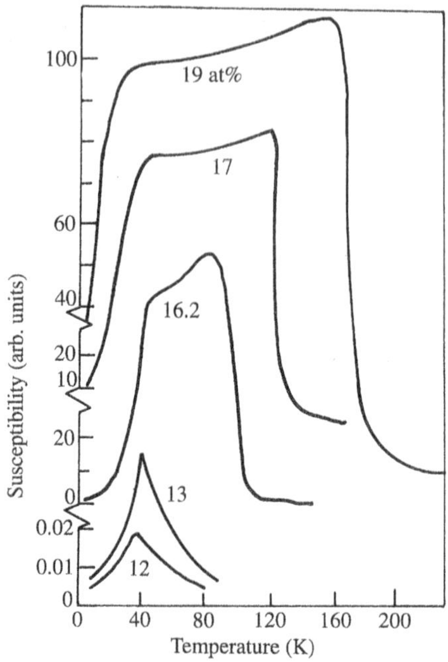

Fig. 6.16 The a.c. susceptibilities (~40 Am^{-1}, 210 Hz) of Au–Fe alloys. (After Coles *et al.* 1978.)

lower temperatures below the glass temperature. Whether this affects the uniform component of the spontaneous magnetization is not clear. It is true that the integrated Bragg intensities increase monotonically with decreasing temperature, albeit with some anomalies, but the resolution is not sufficient to distinguish long range correlations from the intrinsic Bragg peak. At the same time the small angle scattering does not access the uniform spontaneous moment.

This behaviour of the small angle scattering is also seen near the percolation concentration of 19 at% iron in Cr–Fe alloys (Shapiro *et al.* 1981). Figure 6.17 shows the variation of small angle scattering intensity with temperature for three alloys to the ferromagnetic side of the percolation concentration. The most concentrated (34 at% iron) simply shows critical scattering at the Curie temperature and then as the temperature decreases a decreasing intensity associated with magnon scattering. The other two alloys show an increase in the Fourier components of the correlation with a wavelength of about 150 Å below about 30 K.

Similar features are again seen for the insulating spin glass system, $Eu_CSr_{1-C}S$. Here the critical composition for ferromagnetic long range order is $C = 0.51$. The magnetic phase diagram (Maletta and Felsch 1980) is shown in Fig. 6.18. Powder neutron diffraction patterns are shown in Fig. 6.19 for an $C = 0.52$ alloy. The patterns are shown with that at $T = 4.3$ K, above T_C, subtracted. It can be

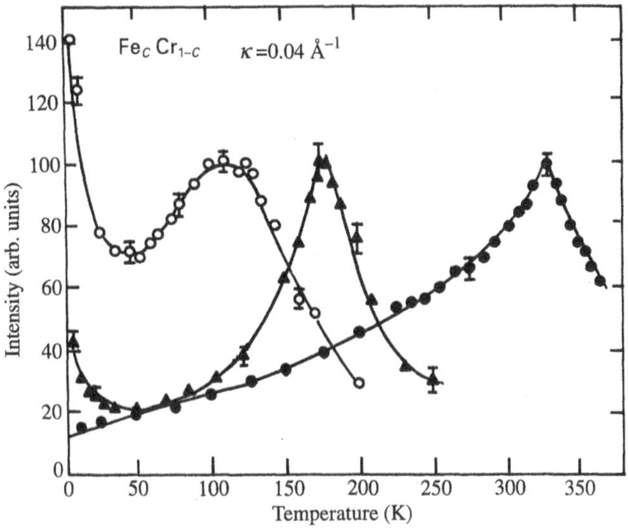

Fig. 6.17 The elastic scattering measured at $\boldsymbol{\kappa} = 0.04$ Å^{-1} for Fe$_C$Cr$_{1-C}$ with $C = 0.22$ (○), 0.26 (▲), and 0.34 (●). (After Shapiro *et al.* 1981.)

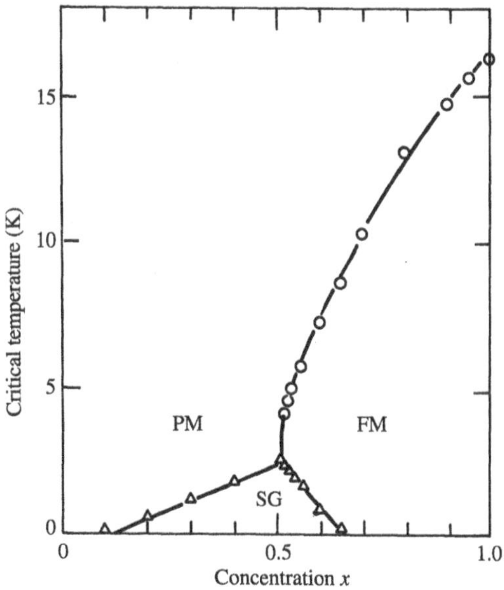

Fig. 6.18 The magnetic phase diagram of Eu$_C$Sr$_{1-C}$S. (After Maletta and Felsch 1980.)

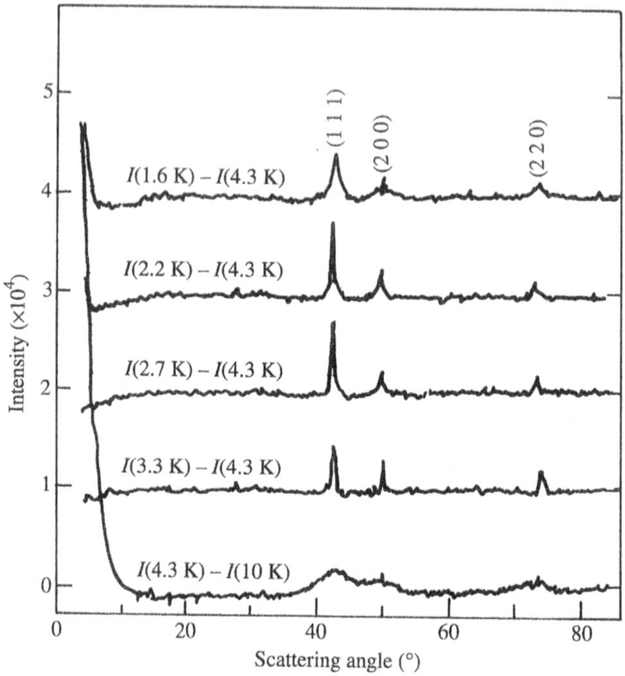

Fig. 6.19 Neutron diffraction patterns of $Eu_{0.52}Sr_{0.48}S$ for various differences in temperature. (After Maletta and Felsch 1980.)

seen that between 10 K and 4.3 K there is some development of magnetic short range order. Then below 4.3 K what appear to be sharp Bragg peaks appear, until at the lowest temperature (T = 1.6 K) a significant broadening of the peaks occurs. It is impossible to say whether a delta function component of the Bragg peaks survives at low temperatures, but if it does its intensity is severely reduced. At the same time the small angle scattering shows features similar to the metallic systems so that below a sharp rise in SANS intensity at T_C the SANS continues to rise with decrease in temperature, quite the opposite to that expected for a normal ferromagnet.

Neutron depolarization is useful to characterize the domain size in ferromagnetic samples. The length scale which is accessible is the time of one Larmor precession times the neutron velocity. These lengths tend to be micrometres rather than hundreds of Ångstroms. The work of Mirebeau *et al.* (1990) with re-entrant Ni–Mn alloys shows another feature of this magnetic state. Mirebeau *et al.* (1990) found that the domain size changed very little at low (560 Am^{-1}) fields on cooling through the re-entrant glass temperature and remained at about 50 μm. At a higher field (9950 Am^{-1}) the difference between the normal ferromagnetic and glassy phase was quite apparent. In the normal phase the domains responded to the increased field and domain size increased to about the thickness of the specimen, 250 μm, whereas the domain configuration was unaltered in the glassy phase. It is clear that here is the main reason for the drop in susceptibility below the glass temperature, but it throws little light on whether the spontaneous ferromagnetic moment does reduce at low temperatures.

Apparently contrasting behaviour is seen for the insulating antiferromagnet $Fe_{0.55}Mg_{0.45}Cl_2$. $FeCl_2$ has an hexagonal structure with ferromagnetically coupled planes which stack antiferromagnetically along the *c*-axis. The magnetism is extremely anisotropic and can be described with an Ising model. The first neighbour, in plane interactions, are ferromagnetic, and the out of plane second neighbour interactions are weaker and antiferromagnetic. The Néel temperature for the alloy investigated by Wong *et al.* (1985) is 7.5 K and freezing behaviour is seen in the a.c. susceptibility below 3 K. The intensity at the peak and off the peak of an antiferromagnetic Bragg reflection is shown in Fig. 6.20(a). The rise in the peak intensity with fall in temperature is smooth and uninterrupted, whilst the scattering just off the peak shows the critical scattering at the Néel temperature. Below the Néel temperature the shorter range correlations persist, but do not increase as they do for the ferromagnetic alloys investigated. More strikingly, there is no evidence for a change in width of the scattering at the Bragg position. Figure 6.20(b) shows a comparison of an antiferromagnetic Bragg peak below the glass temperature and a nuclear Bragg peak. There is no difference in width and the solid lines are fits to resolution limited Gaussians. Figure 6.20(c) shows the diffuse scattering found under antiferromagnetic Bragg peaks at low temperatures. The solid line is a fitted Lorentzian which yields a correlation length of approximately 10 Å. A scan under a nuclear Bragg peak is shown for comparison and to confirm that there is no atomic short range order which could contribute to the diffuse scattering seen under the antiferromagnetic peak. The picture which emerges from this particular study is of a long range ordered antiferromagnetic structure in which some of the spins do not participate, but which freeze into a glassy structure below the freezing temperature.

So the picture of the re-entrant spin glass phase is different for the basically ferromagnetic systems and the basically antiferromagnetic system studied. As the antiferromagnetic system is an Ising system and the ferromagnetic systems Heisenberg in nature, it is not clear whether the difference is due to the magnetic order or to the dimensionality. There are a number of reasons why Ising long range order may be more stable than Heisenberg (Mukamel and Grinstein 1982; Bhatt and Young 1985). On the other hand dipolar interactions tend to inhibit ferromagnetic but not antiferromgnetic order, and this might be crucial in ferromagnets which are close to a critical concentration. This latter possibility is supported weakly in that Mn–Cu, which has a glassy phase re-entrant in antiferromagnetic long range order (Gibbs and Smith 1980), has been well investigated over many years with no reports of anomalous temperature dependence of the antiferromagnetic Bragg peak intensity.

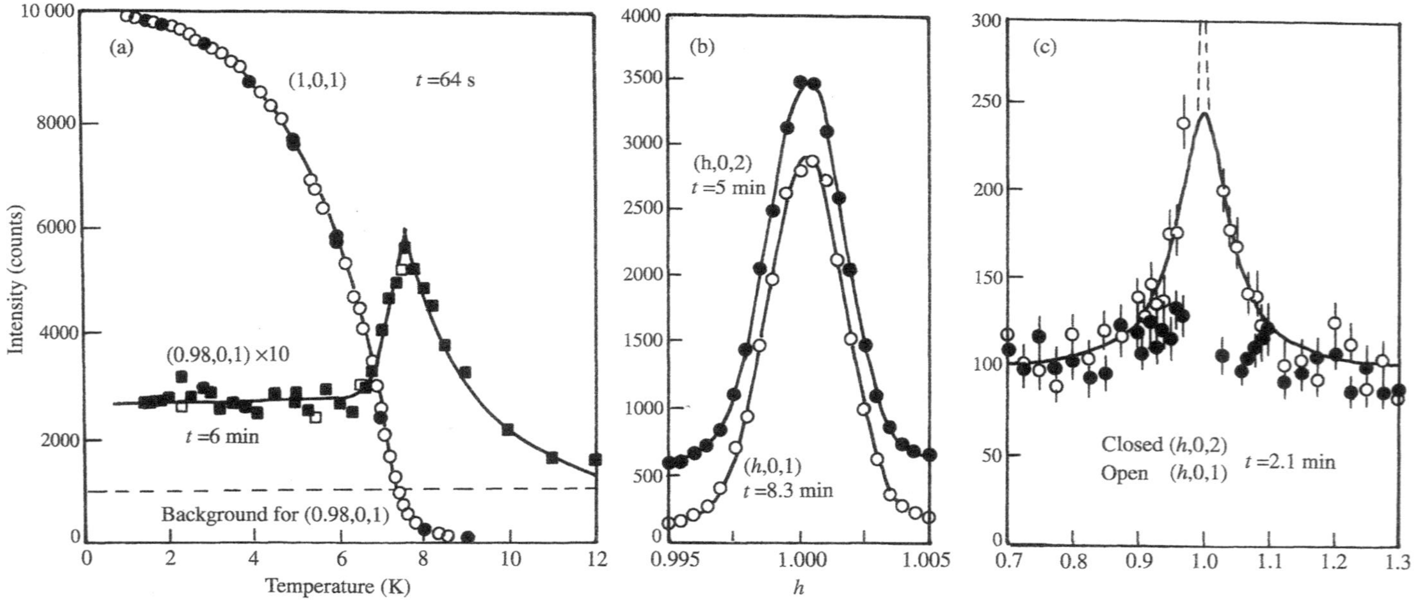

Fig. 6.20 (a) The temperature dependence of the magnetic Bragg scattering and critical scattering, (b) comparison of nuclear and magnetic Bragg peak profiles at 1.30 K, well below the glass temperature, and (c) diffuse scattering near magnetic (○) and nuclear (●) Bragg peaks for $Fe_{0.55}Mg_{0.45}Cl_2$ at T = 1.50 K. In (c) the solid line is a fitted Lorentzian and the dashed line is the Bragg profile. (After Wong *et al.* 1985.)

6.8 SPIN ECHO SPECTROSCOPY

One of the most characteristic properties of spin glasses is that of long time nonexponential relaxation. Conventional triple axis and time of flight spectrometers have only been able to access part of the dynamical behaviour. Much of the insight has come from neutron spin echo spectroscopy with energy resolution of 1 μeV and time resolution of 5×10^{-8} s. Coupled with the resolution is the ability to naturally display the relaxation in the time domain.

Figure 6.21 is a schematic diagram of the IN11 spectrometer at the Institut Laue–Langevin in Grenoble. Neutrons from a cold neutron guide are monochromated by a velocity selector so that the incident wavelength can be changed from 4 Å to about 12 Å. The resulting accessible time domain is from 1.5×10^{-12} s to 4×10^{-8} s and an ultimate energy resolution of 2 neV. A supermirror polarizes the beam after which the polarization is made to precess in the scattering plane around the field in a solenoid leading to the specimen by turning the polarization 90° to the field. On scattering from a paramagnet or a spin glass the component of the polarization parallel to the scattering vector is reversed and the precession direction is thus reversed and the polarization of the scattered beam 'unwinds' in an identical coil between specimen and detector. For elastically scattered neutrons the length needed to regain the initial polarization is the

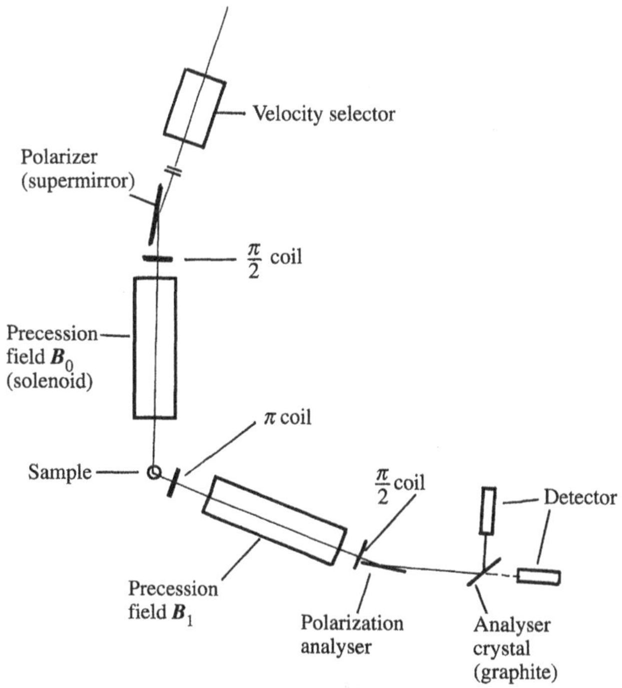

Fig. 6.21 The layout of the IN11 spin echo spectrometer. (After Dagleish *et al.* 1980.)

same as the length on the incident side for identical solenoids. If a 90° polarization turner is located at this point to turn the polarization back into the field direction the polarization is 'recaptured'. All wavelength components in the incident beam undergo different numbers of precessions in the two fields but each wavelength component has an equal number of precessions in the coils before and after the specimen. This is why the spin echo method is capable of a much higher energy resolution than the energy spread in the incident beam.

For small changes in energy on scattering the original polarization is not quite 'recaptured'. The phase change is

$$\phi = 2\pi\Delta n \tag{6.13}$$

where Δn is the change in precession number. The fractional change in energy is related to the fractional change in velocity v which is again directly related to the fractional change in precession number as follows

$$\frac{\Delta E}{E} \approx \frac{2\Delta v}{v} = \frac{2\Delta n}{n}. \tag{6.14}$$

The 'recaptured' polarization is

$$\langle P_x \rangle = P \frac{\int_{-\infty}^{+\infty} \hbar \, d\omega \left(\frac{d^2\sigma}{d\Omega dE} \right) \cos\phi}{\int_{-\infty}^{+\infty} \hbar \, d\omega \left(\frac{d^2\sigma}{d\Omega dE} \right)} \tag{6.15}$$

with

$$\phi = \frac{\pi n \Delta E}{E} = \frac{\pi n \hbar \omega}{E} = \tau\omega. \tag{6.16}$$

The number of precessions in the flight paths n, and therefore the effective time of the measurement τ, is controlled by the magnitude of the guide fields in the flight paths. A series of measurements might entail a survey of $\langle P_x \rangle$ with solenoid field and yield a profile of the specimen's time relaxation.

6.9 SPIN GLASS DYNAMICS

Neutron spin echo results can be related to magnetic susceptibilities in the following way. Expressing the cross section in terms of the generalized susceptibility (1.37) in the quasielastic approximation we have

$$\langle P_x \rangle = P \frac{\int \hbar \, d\omega \left(\frac{e\gamma}{\hbar c} \right)^2 \frac{V}{\pi\beta\hbar\omega} \{ \chi''_{xx}(\kappa,\omega) + \chi''_{yy}(\kappa,\omega) \} \cos\tau\omega}{\int \hbar \, d\omega \left(\frac{e\gamma}{\hbar c} \right)^2 \frac{V}{\pi\beta\hbar\omega} \{ \chi''_{xx}(\kappa,\omega) + \chi''_{yy}(\boldsymbol{\kappa},\omega) \}} \tag{6.17}$$

with $\tau = (\pi n\hbar)/E_0$ and for $\hbar\omega << 1/\beta$ and $k' \approx k$, i.e. for small energy changes. This is justified because of the small energy changes in spin echo experiments. Then

$$\langle P_x \rangle = P \frac{\int_0^{\infty} \mathrm{d}t \{\mathrm{X}_{xx}(\boldsymbol{\kappa},t) + \mathrm{X}_{yy}(\boldsymbol{\kappa},t)\} \int_{-\infty}^{+\infty} \mathrm{d}\omega \frac{\sin\omega t}{\omega} \cos\omega\tau}{\pi \int_0^{\infty} \mathrm{d}t \{\mathrm{X}_{xx}(\boldsymbol{\kappa},t) + \mathrm{X}_{yy}(\boldsymbol{\kappa},t)\}} \tag{6.18}$$

using the Kramers–Krönig relationships. Now

$$\sin\omega t \cos\tau\omega = ½\{\sin\omega(\tau + t) - \sin\omega(\tau - t)\} \tag{6.19}$$

and in the range $0 \le t < \infty$, the left argument is always positive, but for $\tau \le t < \infty$ the right argument is negative. Thus, for $t < \tau$ the integral over ω is 0, but for $t > \tau$ the integral over ω is π. Thus

$$\begin{aligned} \langle P_x \rangle &= P \frac{\int_{\tau}^{\infty} \mathrm{d}t \{\mathrm{X}_{xx}(\boldsymbol{\kappa},t) + \mathrm{X}_{yy}(\boldsymbol{\kappa},t)\}}{\int_0^{\infty} \mathrm{d}t \{\mathrm{X}_{xx}(\boldsymbol{\kappa},t) + \mathrm{X}_{yy}(\boldsymbol{\kappa},t)\}} \\ &= P \left[1 - \frac{\int_0^{\tau} \mathrm{d}t \{\mathrm{X}_{xx}(\boldsymbol{\kappa},t) + \mathrm{X}_{yy}(\boldsymbol{\kappa},t)\}}{\int_0^{\infty} \mathrm{d}t \{\mathrm{X}_{xx}(\boldsymbol{\kappa},x) + \mathrm{X}_{yy}(\boldsymbol{\kappa},t)\}} \right] \end{aligned} \tag{6.20}$$

and it can be seen that the spin echo experiment measures the fraction of spins which have participated in the relaxation process after a time $\tau = (\pi n\hbar)/E_0$; specifically the magnetization induced after time τ per unit applied field. Now

$$\int_0^{\infty} \mathrm{d}t \{\mathrm{X}_{xx}(\boldsymbol{\kappa},t) + \mathrm{X}_{yy}(\boldsymbol{\kappa},t)\} = \chi'_{xx}(\boldsymbol{\kappa}, \omega = 0) + \chi'_{yy}(\boldsymbol{\kappa}, \omega = 0) \tag{6.21}$$

and the energy integrated magnetic cross section is

$$\frac{\mathrm{d}\sigma}{\mathrm{d}\Omega} = 2V \left(\frac{e\gamma}{\hbar c} \right)^2 k_B T \; \{\chi'_{xx}(\boldsymbol{\kappa},0) + \chi'_{yy}(\boldsymbol{\kappa},0)\} \tag{6.22}$$

so that the susceptibility induced after $t = \tau$ can be obtained from a NSE experiment coupled with an absolute measurement of the energy integrated magnetic cross section. Both can be done on a NSE spectrometer.

Neutron spin echo results on copper 5 at% manganese have been discussed previously. Now it can been explicitly seen how the estimates of of spin echo results at longer times could be obtained from a.c. susceptibility results for comparison (Fig. 6.9). Strictly one needs the a.c. susceptibility for all frequencies to obtain the susceptibility induced after a particular time τ, but rough estimates can be made by taking the a.c. susceptibility with frequency $\omega = 2\pi/\tau$. Alternatively one can obtain estimates of the a.c. susceptibility from NSE measurements by the reverse procedure. This permits the a.c. susceptibility to be followed in frequency ranges inaccessible to any other technique. Figure 6.22 shows susceptibilities estimated in this way for copper 5 at% manganese (Mezei 1982) as a function of temperature and are labelled according to frequency. Common to all glass systems is the shift in the susceptibility maximum to lower temperatures for longer times.

The NSE results clearly show a change in the form of the relaxation from close to exponential at high temperatures to a –ln t variation at lower temperatures. The broad features of this behaviour can be fitted (Mezei 1982) using a simple superposition of Arrhenius laws so that

$$\frac{\int_0^\tau \mathrm{d}t\, X(\boldsymbol{\kappa},t)}{\int_0^\infty \mathrm{d}t\, X(\boldsymbol{\kappa},t)} = 1 - \frac{1}{E_{\max}} \int_0^{E_{\max}} \mathrm{d}E \; \mathrm{e}^{-\tau/\tau_0 \exp(E/k_B T)}. \tag{6.23}$$

This assumes that the average contribution to the susceptibility of relaxing entities is the same for all activation energies. This means that all sizes of spin cluster must have a similar distribution of activation energies, which means that the simple application of the model requires little change of the relaxation spec-

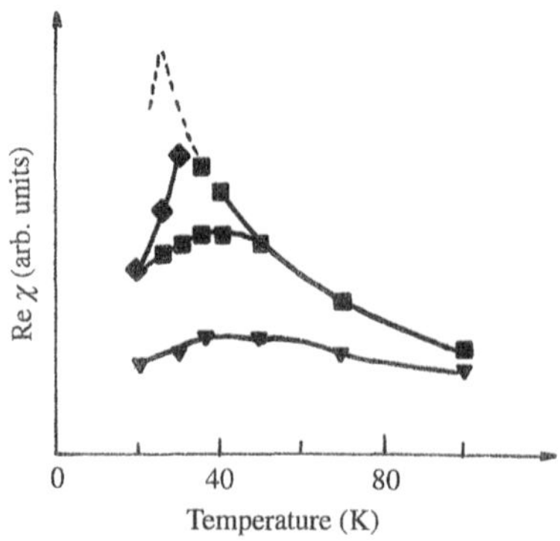

Fig. 6.22 The temperature dependence of the high frequency susceptibility for Cu–Mn 5 at% calculated from spin echo data. (----) 10 Hz, (◆) 100 MHz, (■) 2.5 GHz, (▼) 40 GHz. (After Mezei 1982.)

trum with **κ**. This is certainly true for all the spin glasses examined by NSE except close to the glass temperature and when there are very strong spin correlations (Mezei 1982). Figure 6.23 shows the results of fitting $Eu_{0.4}Sr_{0.6}S$ NSE spectral data with expression (6.23) and with E_{max} = 13 K and $\tau_0 = 3.7 \times 10^{-11}$ s. The glass temperature at long times is 1.6 K, and as can be seen the points are quite well described over two decades of time and for temperatures from 12 to 2.5 K. At lower temperatures the relaxation is slower than would be predicted, and this corresponds to a hardening of the spectrum of activation energies. In some sense this is an indication of a cooperative behaviour near the glass temperature.

Inelastic scattering experiments involving other types of instruments (time of flight or triple axis intruments) support the picture of spin glass dynamics revealed by NSE, that is, a broad spectrum of excitation energies with very little **κ** dependence. Studies at higher **κ** like Scheuer *et al.*'s (1989) have revealed some structure in the energy spectrum of Au–Fe due to isolated iron atoms, but in particular, for simple spin glasses, no coherent excitations. This however is not true for materials with a high degree of spin correlation, such as (Cd,Mn)Te. Figure 6.24 shows the broad magnon groups observed by Giebultowicz *et al.* (1989) in $Cd_{0.33}Mn_{0.67}Te$. Another material showing sharp excitations is the Ising spin glass $Fe_{0.4}Mg_{0.6}Cl_2$. The longitudinal and transverse fluctuations in this system are quite different. Wong *et al.* (1987) found that with a triple axis spectrometer resolution of 100 μeV they could not resolve the energy spread of the longitudinal fluctuations. However the transverse fluctuations had energies of more than 1 meV. The transverse fluctuations showed some dispersion in the plane of the layered structure but can essentially be described as excitations near

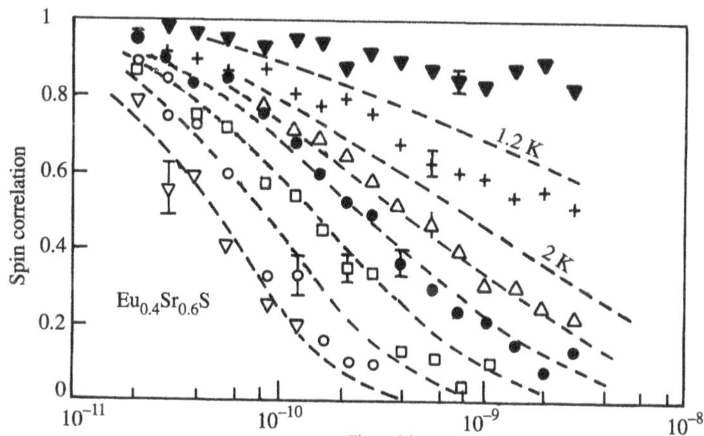

Fig. 6.23 Relaxation of the spin correlations at **κ** = 0.072 Å^{-1} from spin echo measurements. The dashed lines are from eqn (6.23) E_{max}=13 K and $\tau_0 = 3.7 \times 10^{-11}$ s, (▼)1.2 K, (+) 2 K, (Δ) 2.5 K, (●) 3 K, (□) 4 K, (○) 6 K, (∇) 12 K. (After Mezei *et al.* 1988.)

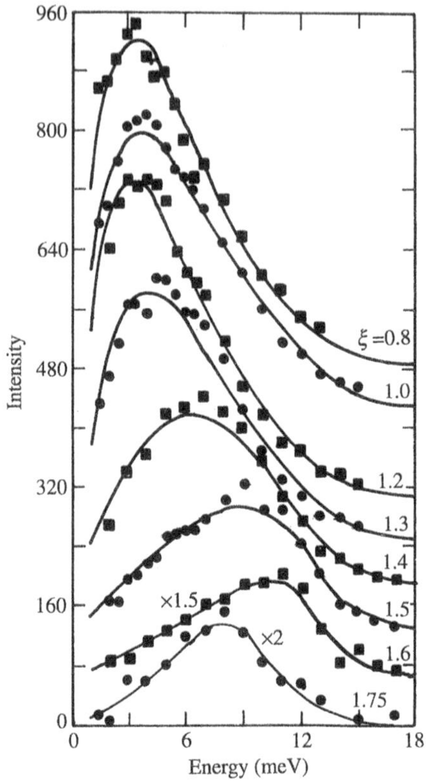

Fig. 6.24 Inelastic scattering from $Cd_{0.33}Mn_{0.67}Te$ at 4.2 K along the $[1\ \bar{1}\ 0]$ direction, $\boldsymbol{\kappa} = 2\pi/a(\zeta,\zeta,0)$. The data are progressively displaced upwards. (After Giebultowicz *et al.* 1989.)

the zone centre with an energy corresponding to the single ion anisotropy, and broad excitations near the zone boundary of a higher energy, probably reflecting the average exchange energy. Between these two extremes the excitations are too broad to be easily resolved. The difference in the dynamic behaviour of the longitudinal and transverse fluctuations is also seen in the a.c. susceptibility where the longitudinal susceptibility shows a freezing temperature and the transverse susceptibility is featureless.

6.10 MAGNONS IN RE-ENTRANT SPIN GLASSES

Magnons are quite clearly in evidence in ferromagnetic re-entrant spin glasses. Murani (1983) investigated gold 19 at% iron using time of flight spectroscopy and found inelastic peaks resolved from the central peak. Figure 6.25 shows the results for temperatures in the ferromagnetic phase and for 70 K and 35 K in the re-entrant phase. The solid lines in the diagram are the results of fitting the

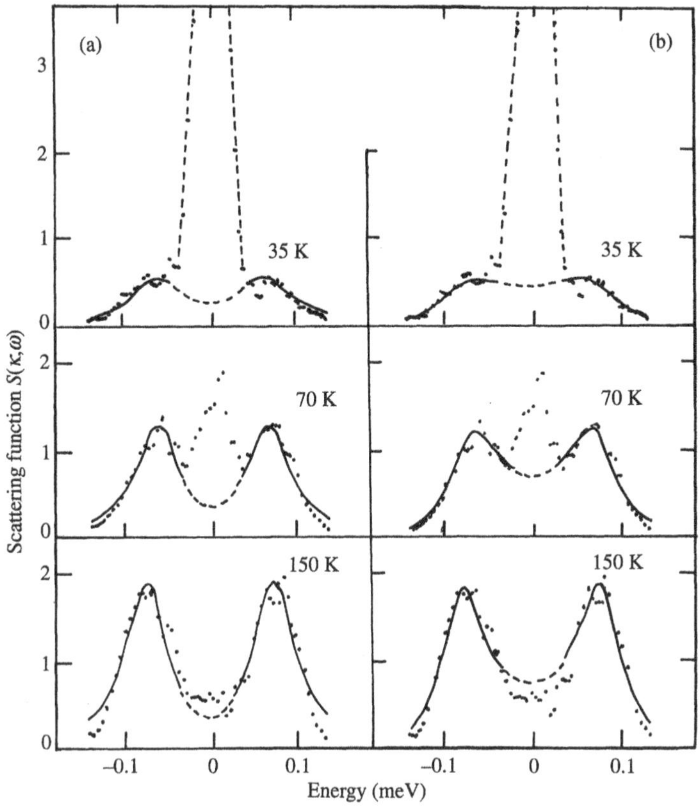

Fig. 6.25 Inelastic scattering from Au–Fe 19 at% at various temperatures and κ=0.055 Å^{-1}. The lines are fits of (a) a double peaked Lorentzian and (b) a damped harmonic oscillator. (After Murani 1983.)

magnons with a double Lorentzian (DL) spectral function or a damped harmonic oscillator (DHO) spectral function. Neither can be favoured because the physics of the magnon broadening is not known, but Murani (1983) found that the DHO was a better fit to the data. However, the interpretation of the two fits is quite different. The DL fit yields magnon energies which decrease substantially at low temperatures whereas for the DHO fit magnon energies are constant at low temperatures. The former suggests that the ferromagnetism is weakening at low temperatures. At the same time, at temperatures below 70 K a strong quasielastic peak grows in intensity. Whether or not the ferromagnetism is weakening at low temperatures, the observation of magnons with a wavelength of 114 Å suggests that there are significant regions in which there is ferromagnetic correlation over more than this length.

Another careful and detailed experiment was done by Hennion *et al.* (1984) on the ferromagnetic re-entrant system Ni–Mn. The specimen was polycrystalline and cooled rapidly from high temperatures to induce as high a degree of randomness in the distribution of manganese and nickel atoms as possible. Ni–Mn alloys close to the composition Ni_3Mn tend to order in the Cu_3Au structure and this ordering has a profound effect on the magnetic properties. The 21.6 at% manganese alloy used in this study had a Curie temperature of 330 K and a glass temperature of about 40 K. The cold neutron triple axis measurements were corrected for background and nuclear scattering to isolate the magnetic scattering which is shown in Fig. 6.26. The absence of a central peak in the high temperature results is circumstantial evidence for the correctness of the background corrections. The solid lines in Fig. 6.26 are fits to the DHO model with a Lorentzian quasielastic peak folded with the instrument resolution, but fits were also obtained using the DL model. Again the two models yield different variations with temperature for the magnon energies with the latter showing a reduction in energy at low temperatures. In addition, both models yield a strong increase in magnon energy below about 70 K, and a strong quasielastic peak appears below 100 K.

It is clear that the ferromagnetic magnons persist into the re-entrant glass region. Hennion *et al.* (1984) state that only instrumental limitations stopped the observation of magnons at lower temperatures. The magnon widths show some increase at high temperatures, as would be expected, but the magnons also

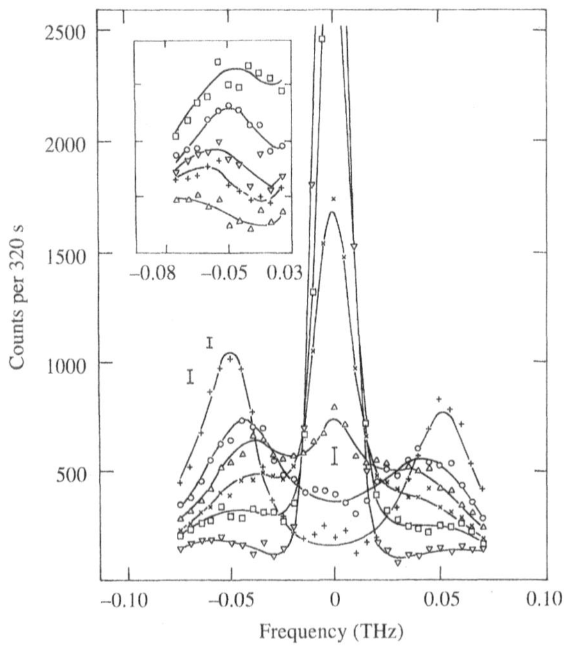

Fig. 6.26 Magnetic scattering at various temperatures from $Ni_{0.784}Mn_{0.216}$ obtained by correcting for nuclear diffuse scattering and background. In the main diagram the temperatures are 180 K (+), 120 K (○), 98 K (Δ), 78 K (×), 58 K (□) 38 K (∇). In the inset 58 K (□), 48 K (○), 38 K (∇), 28 K (+), 16 K (Δ). (After Hennion *et al.* 1984.)

broaden below about 100 K before narrowing again at lower temperatures for **κ** less than 0.055 Å^{-1}. The smallest **κ** value examined was 0.04 Å^{-1}, again implying that ferromagnetic correlation is longer than 155 Å. The transverse static susceptibility derived from integration over the fitted magnons rises with temperature decrease and shows evidence of an extra rise between 200 K and 100 K. Thus the behaviour of the magnons is complex apart from the clarity of their existence and the maintenance of their stiffness in the glassy regime. Similar results are obtained in other ferromagnetic re-entrant systems. For instance in Cr–Fe, Lequien *et al.* (1988) find a similar variation of the magnon energy with temperature.

From the Lorentzian fit of the quasielastic peak in Ni–Mn (Hennion *et al.* 1984) demonstrated a linewidth which depends exponentially on temperature. Figure 6.27 is a plot of the width against inverse temperature and the consequent variation of relaxation time with temperature suggests that in the accessible time interval ($t \leq 10^{-9}$s) the relaxation obeys an Ahrrhenius law, although because of the energy resolution this is very difficult to distinguish from a superposition of Ahrrenius laws.

Of the re-entrant ferromagnetic systems which have been extensively studied, only in (Eu,Sr)S has no evidence for magnons been found, either in the nominally ferromagnetic temperature range, or in the glassy region. This is perhaps because in this system the energies are very much smaller than in the metallic systems cited above. Even in the metallic systems the energy resolution of triple axis and time of flight spectroscopy is only just sufficient to be able to follow the

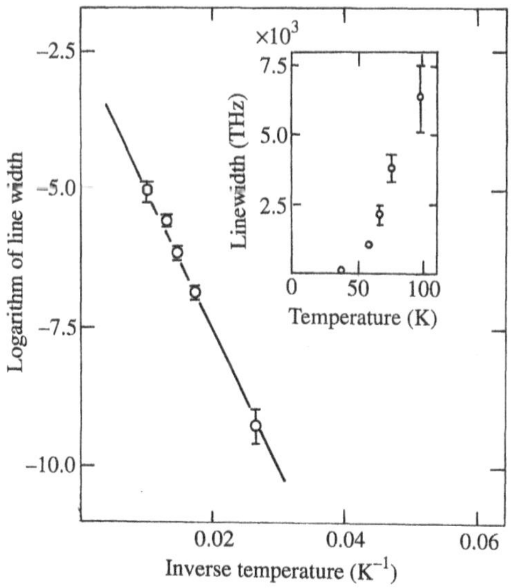

Fig. 6.27 The linewidth of the quasielastic scattering as a function of T (inset). The straight line in the main graph corresponds to a single blocking energy. (After Hennion *et al.* 1984.)

magnons right down into the glassy region. This may be the reason why no magnons are clearly seen in the early data of Maletta *et al.* (1981). However the NSE data of Shapiro *et al.* (1985) also show no evidence for propagating modes. Magnons should be visible in NSE spectroscopy as an oscillation of the time-dependent correlation function. The data for $Eu_{0.54}Sr_{0.46}S$ (Fig.6.28) show no oscillations. If, in the ferromagnetic region at $T = 3.2$ K, the only magnetic excitations were magnons as for the metallic systems, the oscillation should affect the whole of the correlation function, but all that is seen at this temperature is a relaxation which is between logarithmic in time and an exponential decay. At lower temperatures the oscillation might be masked by the quasielastic scattering, but not in the ferromagnetic regime. This raises the question of whether $Eu_{0.54}Sr_{0.46}S$ is different to the metallic ferromagnets showing re-entrant spin glass behaviour.

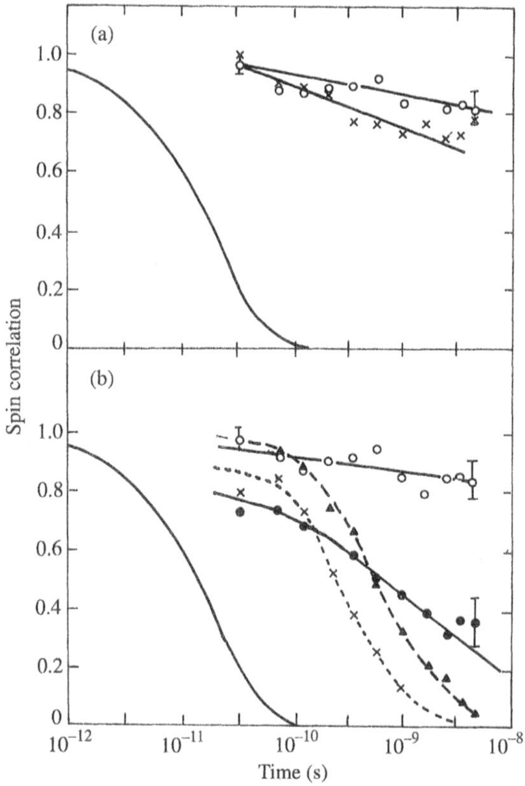

Fig. 6.28 The real time spin relaxation for $Eu_{0.54}Sr_{0.46}S$ measured by spin echo. (a) The κ dependence at $T = 1.2$ K ($T_g = 2.0$ K), $e^{-\gamma t}$ ($\gamma = 33$ uV), (○) $\kappa = 0.072$ Å^{-1}, (×) $\kappa = 0.108$ Å^{-1}. (b) the t dependence for $\kappa = 0.72$ Å^{-1}(○) $T = 1.2$ K, (●) $T = 3.2$ K, (▲) $T = 5.0$ K, (×) $T = 6.0$ K. (After Shapiro *et al.* 1985.)

Neutron scattering has been crucial for the understanding we have of spin glasses. Without neutron scattering we would be hard pressed to distinguish them from antiferromagnets. The exact nature of the re-entrant spin glass phase has yet to be established and undoubtedly neutron scattering will have a large role to play. The nature of the spin glass transition also needs more attention but it may be that in this area neutron scattering will be less useful as the critical behaviour of spin glasses appears to more usefully described in terms of the non-linear part of the susceptibility rather than the susceptibility itself.

References

Ahmed, N. and Hicks, T.J. (1974). *Solid State Comm.* **15**, 415.

Alloul, H., Hennion, B., and Mendels, P. (1988). *J. de Phys.* **C8**, 1019.

Arrott, A. (1965). *J. Appl. Phys.* **36**, 1093.

Bacon, G.E., Dunmur, I.W., Smith, J.H, and Street, R. (1957). *Proc. Roy. Soc.* **A241**, 223.

Bhatt, R.N. and Young, A.P. (1985). *Phys. Rev. Lett.* **54**, 924.

Blume, M. (1963). *Phys. Rev.* **130**, 1670.

Cable, J.W., Werner, S.A., Felcher, G., and Wakabyashi, N. (1984). *Phys. Rev. B* **29**, 1268.

Cable, J.W., Parette, G., and Tsunoda, Y. (1987). *Phys. Rev. B* **36**, 8467.

Coles, B.R., Sarkissian, B.V.B., and Taylor, R.H. (1978). *Phil. Mag.* **B37**, 489.

Dagleish, P.A., Hayter, J.B., and Mezei, F. (1980). *Lecture Notes in Physics* 128, p. 66. Springer, Berlin.

Davis, J.R. and Hicks, T.J. (1979). *J. Phys. F* **9**, 753.

Davis, J.R., Burke, S.K., and Rainford, B.D. (1980). *J. Mag. Mag. Mat.* **15–18**, 151.

Edwards, S.F. and Anderson, P.W. (1975). *J. Phys. F* **5**, 965.

Gibbs, P. and Smith, J.H. (1980). *J. Mag. Mag. Mat.* **15–18**, 155.

Giebultowicz, T. and Holden, T.M. (1988). *Semiconductors and Semimetals* **25**, 125.

Giebultowicz, T., Rhyne, J.J., Furdyna, J.K., and Galazka, R.R. (1988). *J. de Phys.* **C8**, 1199.

Giebultowicz, T., Rhyne, J.J., Ching, W.Y., Huber, D.L., Furdyna, J.K., Lebech, B. *et al.* (1989). *Phys. Rev. B*

Gray, E.M., Hicks, T.J., and Smith, J.H. (1982). *J. Phys. F*

Harders, T.M. and Wells, P. (1983). *J. Phys. F* **13**, 1017.

Harders, T.M., Hicks, T.J. and Smith, J.H. (1983). *J. Phys. F*

Hennion, B., Hennion, M., Hippert, F., and Murani, A.P. (1984). *J. Phys. F* **14**, 489.

Kasuya, T. (1956). *Prog. Theor. Phys.* **16**, 45.

Kasuya, T. (1970). *IBM J. Res. Dev.* **14**, 214.

Kouvel, J.S. (1961). *J. Phys. Chem. Solids* **21**, 57.

Lequien, S., Hennion, B., and Shapiro, S.M. (1988). *Phys. Rev. B* **38**, 2669.

Maletta, H. and Felsch, W. (1978). *J. de Phys.* **C6**, 931.

Maletta, H. and Felsch, W. (1979). *Phys. Rev. B* **20**, 1245.

Maletta, H. and Felsch, W. (1980). *Z. Physik B* **37**, 55.

Maletta, H., Zinn, W., Scheuer, H., and Shapiro, S.M. (1981). *J. Appl. Phys.* **52**, 1735.

Meneghetti, D. and Sidhu, S.S. (1957). *Phys. Rev.* **105**, 130.

Mezei, F. (1982) *J. Appl. Phys.* **53**, 7654.

Mezei, F. and Murani, A.P. (1979). *J. Mag. Mag. Mat.* **14**, 211.

Mezei, F., Maletta, H., Farago, B., and Shapiro, S.M. (1988). *J. de Phys.* **C8**, 1045.

Mirebeau, I., Itoh, S., Watanabe, N., Endoh, Y., Hennion, M., and Calmettes, P. (1990). *J. Appl. Phys.* **67**, 5232.
Morgownik, A.F.J. and Mydosh, J.A. (1981). *Physica* **107B**, 305.
Mukamel, D. and Grinstein, G. (1982). *Phys. Rev. B* **25, 381.**
Murani, A. (1976). *Phys. Rev. Lett.* **37**, 450.
Murani, A. (1980). *Solid State Comm.* **34**, 705.
Murani, A.P. (1983). *Phys. Rev. B* **28**, 432.
Murani, A. and Tholence, J.L. (1977). *Solid State Comm.* **22**, 25.
Owen, J., Browne, M.E., Arp, V., and Kip, A.F. (1957). *J. Phys. Chem. Solids* **2**, 85.
Ruderman, M.A. and Kittel, C. (1954). *Phys. Rev.* **96**, 99.
Sato, H., Werner, S.A. and Kikuchi, R. (1974). *J. de Phys.* **C4**, 23.
Scheuer, H., Löwenhaupt, M., and Just, W. (1989). Preprint
Shapiro, S.M., Fincher, C.R., Palumbo, A.C., and Parks, R.D. (1981). *Phys. Rev. B* **24**, 6661.
Shapiro, S.M., Maletta, H., and Mezei, F. (1985). *J. Appl. Phys.* **57**, 3485.
Soukoulis, C.M., Grest, G.S., and Levin, K. (1978). *Phys. Rev. Lett.* **41**, 568.
Steigenberger, U. and Lindley, L. (1988). *J. Phys. C* **21**, 1703.
Steigenberger, U., Lebech, B., and Galazka, R.R. (1986). *J. Mag. Mag. Mat.* **54–57**, 1285.
Tsunoda, Y. (1989). *J.Phys.:Condens. Matter* **1**, 10427.
Wells, P. and Smith, J.H. (1971) *J. Phys. F* **1**, 763.
Werner, S.A. and Cable, J.W. (1981) *J. Appl. Phys.* **52**, 1757.
Wong, Po-zen, von Molnar, S., Palstra, T.T.M., Mydosh, J.A., Yoshizawa, H., Shapiro, S.M. *et al.* (1985). *Phys. Rev. Lett.* **55**, 2043.
Wong, Po-zen, Yoshizawa, H., Shapiro, S.M., Mydosh, J.A., van den Berg, J., and Palstra, T.T.M. (1987). *Phys. Rev. Lett.* **58**, 1276.
Yosida, K. (1957). *Phys. Rev.* **106**, 893.

7

Amorphous magnets

Atomically glassy alloys are one further step away from regularity in materials. The magnetic systems which have been discussed are at least regular in the underlying lattice. The randomness has been restricted to the occupancy of the sites of the regular lattice. Amorphous materials have no regular lattice although it must be said that all practical amorphous materials show strong preferences for particular atomic distances. In terms of diffraction patterns this means that although there are no Bragg peaks the scattering is strongly structured.

The simplest of magnetic amorphous alloys are ferromagnetic. Naively it might be expected that this would mean that that all atomic magnetic moments would be parallel, and that is what the original analyses of ferromagnetic amorphous diffraction patterns assumed. It has been only with the advent of neutron polarization analysis that it has been shown that this may not be so. However, we start by discussing experiments in which the collinear assumption was made.

7.1 THE STRUCTURE OF FERROMAGNETIC GLASSES

There are a number of ways of separating the atomic and magnetic structures of ferromagnetic glassy materials. Some methods, like the comparison between high and low temperature neutron diffraction patterns, which involve switching a magnetic field along the scattering vector in a neutron experiment, and comparing neutron and X-ray diffraction patterns, do not provide information about the separate correlations between the different types of atoms. In the special case of an amorphous alloy consisting of one magnetic and one non-magnetic atom species this information can be obtained using polarized neutron diffraction.

For each state of the incident polarization the cross section is entirely diffuse and consists of a coherent part arising from the scattering lengths and average moments of the constituents, a Laue part, a truly incoherent part coming from the isotopic and spin incoherent cross sections of the constituents, and a possible contribution from deviations from the average of the moments of the constituent atoms. In addition there is multiple scattering.

There are three independent incident polarization conditions. With the magnetic field along the scattering vector the magnetic scattering is extinguished. With the magnetic field perpendicular to the scattering vector there are two different cross sections for the polarization along and opposite to the field. For a two component amorphous material the three partial structure factors can be obtained from the three different cross sections. For the field parallel to the scattering vector

$$\left(\frac{d\sigma^{+}_{coh}}{d\Omega}\right)_{\parallel} = \left(\frac{d\sigma^{-}_{coh}}{d\Omega}\right)_{\parallel} = C_1^2 b_1^2 \, F_{11}(\kappa) + 2C_1C_2b_1b_2F_{12}(\kappa) + C_2^2 b_2^2 \, F_{22}(\kappa) \qquad (7.1)$$

where the two signs are for the two incident polarizations. For the field perpendicular to the scattering vector

$$\left(\frac{d\sigma^{\pm}_{coh}}{d\Omega}\right)_{\perp} = C_1^2 (b_1 \pm p_1)^2 F_{11}(\kappa) + 2C_1C_2b_2(b_1 \pm p_1)F_{12}(\kappa) + C_2^2 b_2^2 \, F_{22}(\kappa). \qquad (7.2)$$

The equations are simplified by writing in a magnetic moment on only one species. This is all that is necessary for a large number of glassy ferromagnets. The magnetic scattering length is

$$p_1 = \left(\frac{e^2\gamma}{2m_e c^2}\right)\mu_1 f(\kappa) \qquad (7.3)$$

and the partial structure factor (Faber and Ziman 1965) is given by

$$F_{AB}(\kappa) = 1 + \left(\frac{4\pi}{v}\right)\int_0^\infty \left[P_{AB}(r) - 1\right]\frac{r}{\kappa}\sin(\kappa r)\, dr \qquad (7.4)$$

in which $P_{AB}(r)$ is the probability per unit volume of finding an A atom at a distance r from a B atom and v is the average volume per atom. This does not include the self-correlation term which is in the Laue cross sections

$$\left(\frac{d\sigma^{\pm}_{Laue}}{d\Omega}\right)_{\parallel} = C_1C_2(b_1 - b_2)^2$$

$$\left(\frac{d\sigma^{\pm}_{Laue}}{d\Omega}\right)_{\perp} = C_1C_2\left[(b_1 \pm p_1) - (b_2 \pm p_2)\right]^2. \qquad (7.5)$$

The truly incoherent term is normally assumed to consist just of the incoherent nuclear cross sections although this might have to be reassessed in the light of experiments to be discussed later (Cowley *et al.* 1988) which indicate significant fluctuation of the moment transverse to the ferromagnetic direction in simple ferromagnetic glasses.

Bletry and Sadoc (1975) were the first to apply the above formalism to the analysis of polarized neutron diffraction from a $Co_{83}P_{17}$ glass. To normalize their data they first used the limiting cross section at large $\boldsymbol{\kappa}$

$$\left(\frac{d\sigma}{d\Omega}\right)_{\kappa\to\infty} = C_1\left(b_1^2 + \sigma_1^{inc}\right) + C_2\left(b_2^2 + \sigma_2^{inc}\right) + \text{multiple scattering} \qquad (7.6)$$

where the the partial structure factors tend to 1, the magnetic form factor to zero, and the scattering becomes $\boldsymbol{\kappa}$ independent. At small $\boldsymbol{\kappa}$ values the incoherent scattering tends to zero leaving the Laue, incoherent and multiple scattering cross sections and providing a check on the previous normalization. The two normalization procedures agreed within 2 per cent and yielded the three cross sections (normalized to unity at large $\boldsymbol{\kappa}$) shown in Fig. 7.1.

The three partial structure factors can be calculated from the cross sections knowing all the scattering lengths and cross sections. The magnetic scattering length presented a problem because although the saturation moment of the sample could be measured and the moment per cobalt atom deduced, it was almost certain that the local cobalt moment would be greater than this due to the presence of a quasiuniform negative conduction electron magnetization as in cobalt metal (Moon 1971). In addition, there was every likelihood that the cobalt magnetic form factor would differ to that in the metal. Bletry and Sadoc (1975) adjusted the moment value and the extent of the form factor to ensure that the

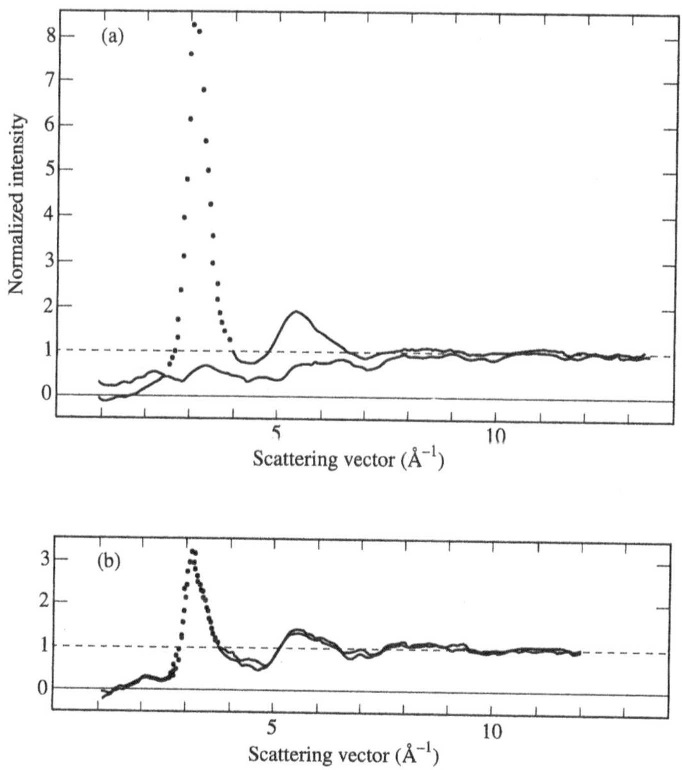

Fig. 7.1 Coherent plus Laue cross sections with the neutron polarization (a) perpendicular, (b) parallel to the scattering vector. The curves have been normalized to one at large κ. (After Bletry and Sadoc 1975.)

partial structure factors oscillated around one in the κ range between 5 $Å^{-1}$ and 10 $Å^{-1}$. This resulted in a local cobalt moment 25 per cent in excess of the average value with a large error and a moment distribution 25 per cent contracted in space than that in the metal with a somewhat smaller error. The resulting partial structure factors after this procedure are shown in Fig. 7.2.

Similar information has been obtained for $Fe_{83}B_{17}$ (Wu *et al.* 1982) except that because of the small concentration and scattering length of ^{11}B, the B–B correlations could not be obtained. The Fe–Fe and Fe–B radial distribution functions show that the first neighbour Fe–B distance is shorter than that for Fe–Fe. This is confirmed by high resolution unpolarized studies (Ze Xianyu *et al.* 1985) in which data was obtained from a scattering vector of 25 $Å^{-1}$ using the KENS spallation source. In this case the resolution of the two distances could only be clearly seen if all the data were used in Fourier transforming from reciprocal to direct space. This study also suggested that there are two closely spaced

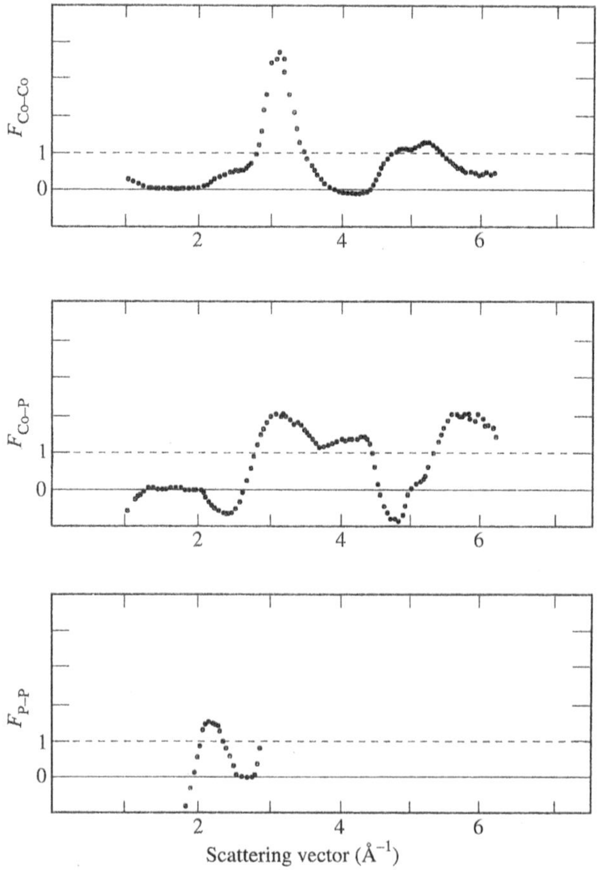

Fig. 7.2 The three partial interference functions. (After Bletry and Sadoc 1975.)

first neighbour iron distances which are not resolved in the study by Wu *et al.* (1982) in which the largest scattering vector was 10 Å^{-1}.

In the simple polarized neutron experiment not only is it assumed that only one species carries a moment but that the moment is not a function of the local environment. The Laue part of the scattering which is subtracted before the coherent partial scattering factors are determined is therefore assumed to have a scattering vector dependence proportional to the square of the magnetic form factor. Fitting of the coherent part also involves both the magnetic moment and the form factor. For FeB a good fit can be obtained if the form factor is that of a more contracted moment distribution of the iron magnetic moment than in metallic iron. All studies agree well on the extent of the contraction; about 80 per cent of that in the metal. Then good agreement is also obtained for the iron moment of 1.75μ_B at room temperature when compared with the moment per iron atom from bulk magnetization measurements.

More recent experiments on FeB using neutron polarization analysis however reveal that the moment is more complicated and has a noncollinear component. Firgure 7.3 shows the results of Cowley *et al.* (1988). The experiment was one in which the sample was in a field of 4 T perpendicular to the scattering vector. Any spin flip scattering is then associated with moment fluctuations which are perpendicular to the polarization (field) direction: that is, they are components not collinear with the average moment direction. It is not that this is a small effect; Cowley *et al.* (1988) estimate that the average canting angle is about 40° which means that if the collinear component is 1.75μ_B the total moment must be in the vicinity of 3μ_B. It is surprising that this extra component was missed in the earlier experiments because although it would not be sensitive to the reversal of neutron polarization it still would have been present in the scattering with field along the scattering vector.

A similar experiment on amorphous Fe–Ni metalloid alloys (Cowley *et al.* 1992) shows similar behaviour. The main difference is the absence of a peak in the spin flip scattering corresponding to the peak in the structure factor (Fig. 7.4) indicating that the noncollinear moment is largely uncorrelated. As the nickel content is increased the noncollinear random component increases. The non spin flip scattering is also larger than expected from the atomic randomness at small scattering vectors which indicates that there is a component from fluctuations in the collinear moment. Both of these indicate that an attractively simple model in which nickel has no moment and that the decrease in moment with increasing nickel concentration is entirely due to the increasing dilution of iron with a constant moment is not correct. Estimates of the average canting angle across the composition range vary from 15 to 30°.

7.2 MAGNONS IN AMORPHOUS FERROMAGNETS

Macroscopically amorphous ferromagnets containing transition elements as their magnetic component do not differ significantly from crystalline ferromagnets

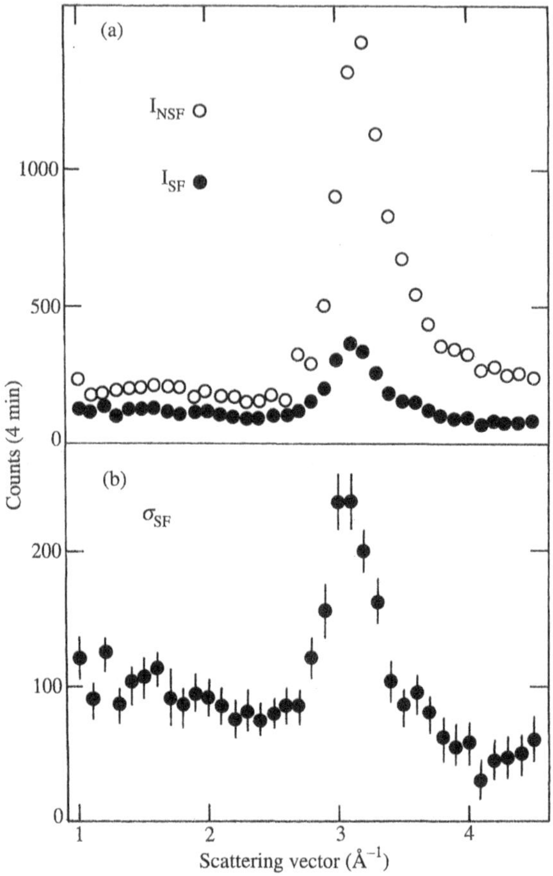

Fig. 7.3 The elastic scattering from $Fe_{83}B_{17}$ in a vertical field of 4 T. (a) (○) I_{NSF}, (●) I_{SF}. (b) (●) σ_{SF}. (After Cowley *et al.* 1988.)

except that anisotropy is small. It is therefore no surprise that at small wavevectors the magnons in glassy ferromagnets differ little from those in crystalline ferromagnets. Typical of the measurements which have been made are those of Mook and Lynn (1984) on metglas ($Fe_{40}Ni_{40}P_{14}B_6$). Figure 7.5 shows the neutron groups associated with magnon creation and annihilation in metglas in triple axis constant q mode. The feature at zero energy is due to the static structure and contains nuclear as well as magnetic scattering. The dispersion relation is the normal quadratic one as shown in Fig. 7.6, and the variation of stiffness with temperature (Fig. 7.7) is as would be expected for a crystalline ferromagnet. It is experimentally difficult in these materials to follow the dispersion relation much beyond those of this study principally because it is steep and there is no chance of observing it other than around the forward scattering direction.

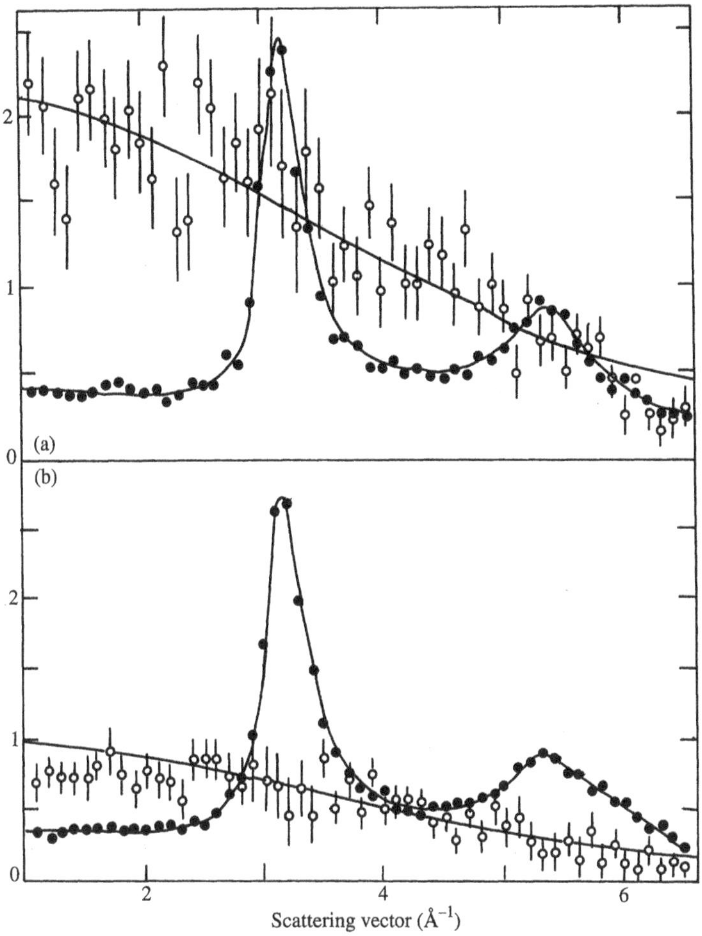

Fig. 7.4 Elastic spin flip (○) × 10 and nonspin flip (●) from Fe_C–Ni_{1-C} amorphous alloys. (a) $C = 0.25$; (b) $C = 0.125$. (After Cowley *et al.* 1992.)

Data confirming particularly the similarity of the low wavevector magnons with those in crystalline materials is to be found in Stusser *et al.* (1979), on a similar glass $(Fe_{65}Ni_{35})_{75}P_{16}B_6Al_3$. Figure. 7.8 shows the magnon stiffness plotted against $(T/T_c)^{5/2}$ according to the Dyson low temperature renormalization prediction. Near the critical temperature the magnon stiffness also follows a power law variation in $(T_c - T)/T$. Figure 7.9 shows that the magnon broadening with scattering vector is also well described by a simple magnon–magnon interaction theory. All these results are consistent with theories which are not particularly applicable to glassy ferromagnets.

There has been considerable interest in amorphous Invar systems such as $Fe_{86}B_{14}$. In this system, as for crystalline systems exhibiting the Invar effect, the

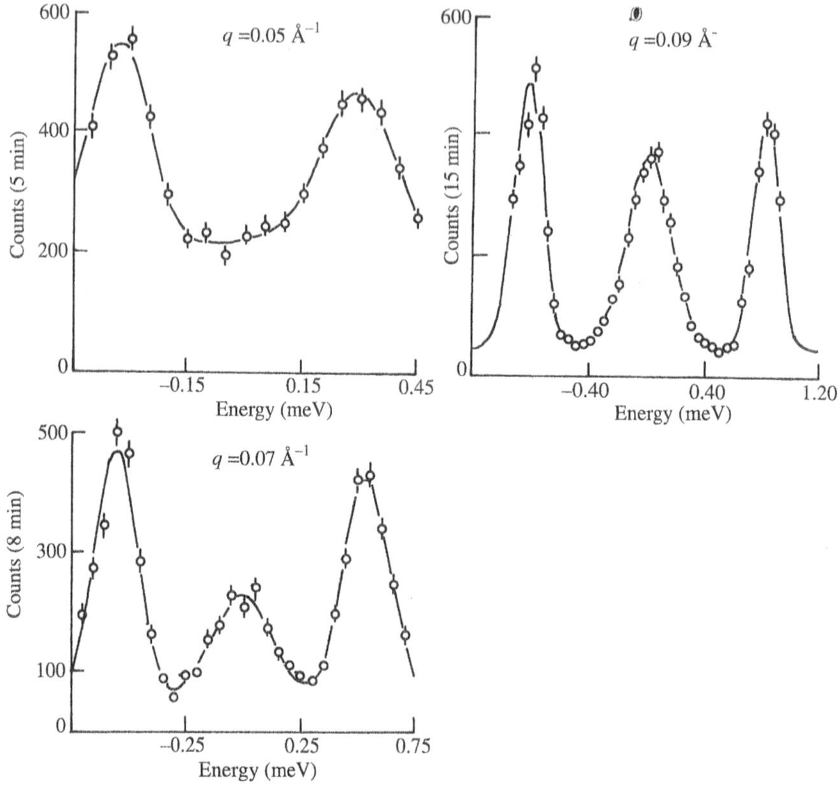

Fig. 7.5 Small wavevector inelastic scattering from $Fe_{40}Ni_{40}P_{14}B_6$ observed at room temperature. (After Mook and Lynn 1984.)

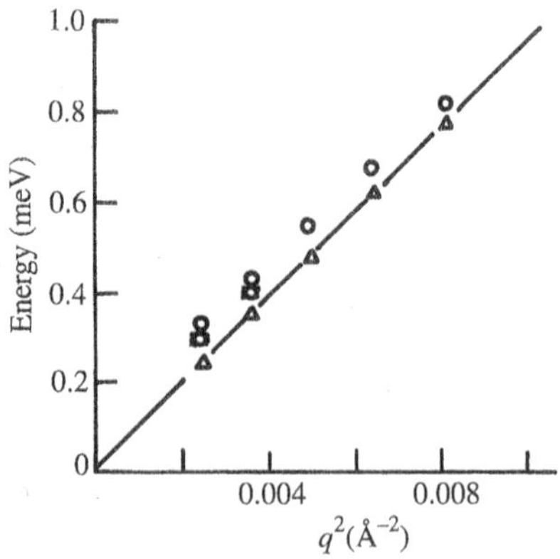

Fig. 7.6 The dispersion relation for energy gain (○) and energy loss (□) in $Fe_{40}Ni_{40}P_{14}B_6$. After correction for instrumental resolution the relation is closely quadratic (Δ). (After Mook and Lynn 1984.)

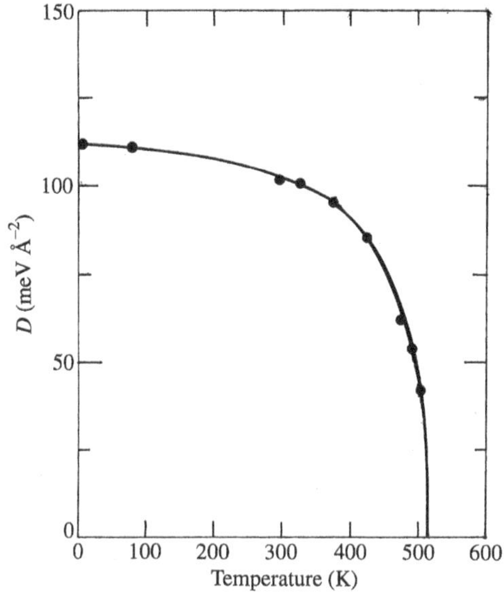

Fig. 7.7 Temperature dependence of the spin wave stiffness constant for $Fe_{40}Ni_{40}P_{14}B_6$. (After Mook and Lynn 1984.)

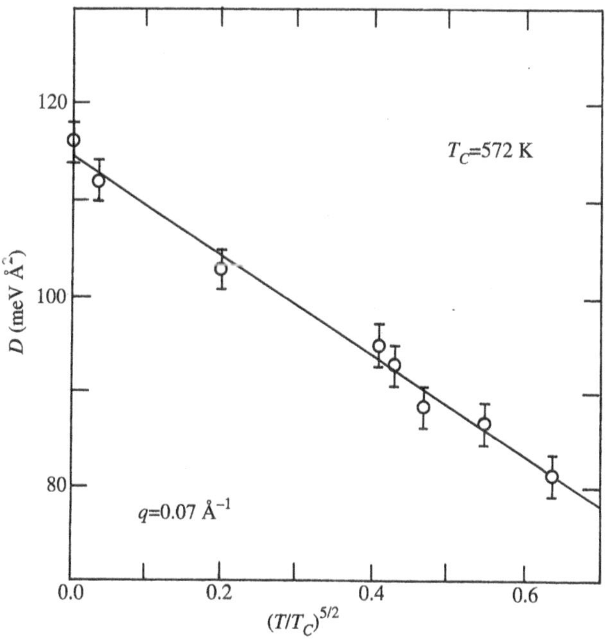

Fig. 7.8 Temperature of the stiffness parameter of $(Fe_{65}Ni_{35})_{75}P_{16}B_6Al_3$. (After Tarvin *et al.* 1978.)

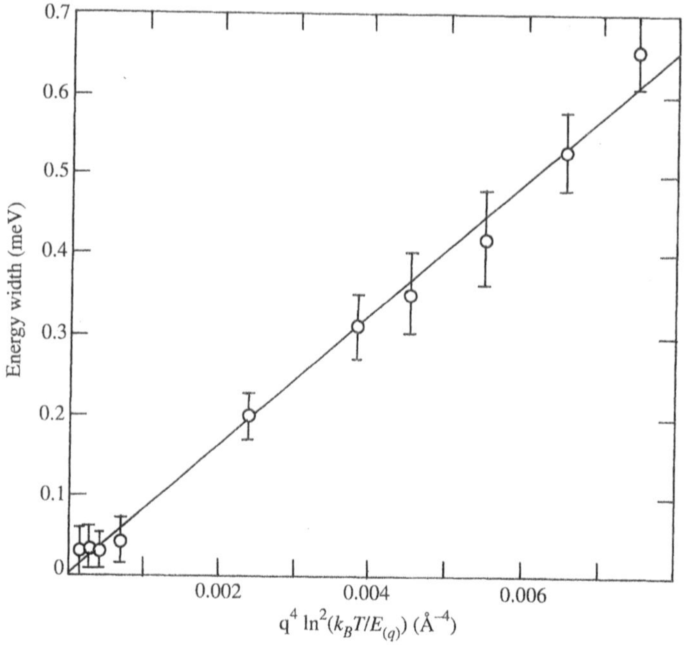

Fig. 7.9 Magnon line widths for $(Fe_{65}Ni_{35})_{75}P_{16}B_6Al_3$ at $T = 450$ K. (After Tarvin *et al.* 1978.)

decrease of magnetization with temperature at low temperatures is faster than would be predicted from the measured spin wave stiffness (Fernandez-Baca *et al.* 1987). However, the measured magnons are wholly unremarkable with the dispersion proportional to q^2 apart from a small energy gap due to dipole–dipole interactions. The magnon linewidths are also well explained by magnon–magnon interactions (Fernandez-Baca *et al.* 1987), contrary to a suggestion by Ishikawa *et al.* (1981). So the behaviour of the magnetization is due to something other than the well-defined magnon modes in these materials.

At larger q the situation is not so clear. There is clearly an increase of magnetic inelastic scattering about the peak in the glass structure factor, at $q{\sim}3$ Å^{-1}. Mook and Lynn (1984) used a polarized neutron time of flight method in which the spectrum is of the difference between the cross section with incident spin up and incident spin down. With the polarization along the scattering vector this eliminates all scattering except that arising from magnetic fluctuations transverse to the ferromagnetic direction (1.38). This, Mook and Lynn (1984) express as the imaginary part of the dynamic susceptibility by dividing the scattering by the Bose factor. The results are shown in Fig. 7.10 for metglas. The contours are of the susceptibility and the dashed lines show the paths traced by the detectors through q–E space. The heart shape may be due to a broadened parabolic dispersion centred at the peak in the structure factor. An earlier experiment on

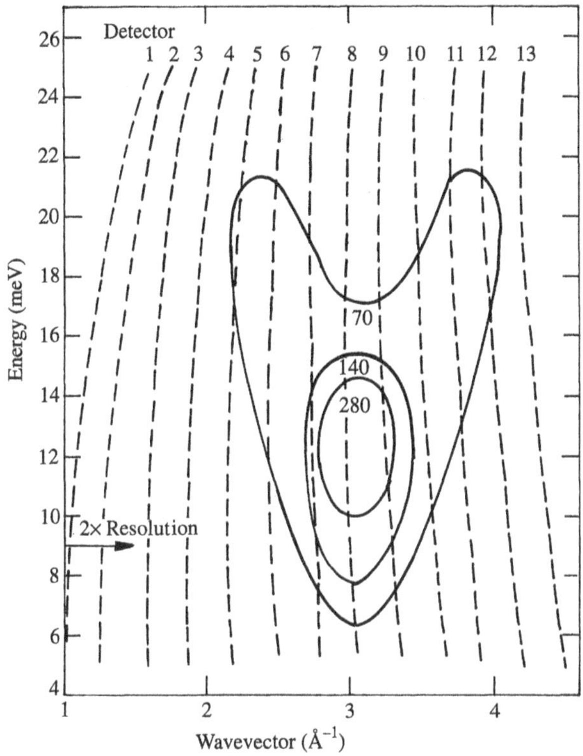

Fig. 7.10 Contour plot of the dynamic susceptibility for $Fe_{40}Ni_{40}P_{14}B_4$. (After Mook and Lynn 1984.)

$Fe_{75}P_{15}C_{10}$ (Mook and Tsuei 1977) shows this more clearly. In all experiments it is clear that the distribution of intensity is very broad with scattering down to very low energy transfers (Shirane *et al.* 1982; Paul *et al.* 1982). The cause is not so clear. The most straightforward explanation is that it is the remains of the magnon dispersion centred at the first reciprocal lattice point of a fully crystalline material which is being seen. However, it is very difficult to calculate this effect in the glassy material but it might be expected that the maximum in the susceptibility might be at a lower energy than that observed.

7.3 AMORPHOUS SPIN GLASSES

It is possible to fabricate amorphous MnSi by sputtering which undergoes a paramagnetic to spin glass transition very similar to that observed for Cu–Mn. At 10 kHz the freezing temperature is 24 K compared with 27 K for 5 at% Cu–Mn. Figure 7.11 shows the the magnetic cross section measured at two temperatures below the glass temperature by Aeppli *et al.* (1985) and isolated by switching the polarization from parallel to perpendicular to the scattering vector in a polarization analysis experiment. There is considerable ferromagnetic correlation and the solid line is a Lorentzian. Aeppli *et al.* (1985) also demonstrated the divergence of the laboratory measured susceptibility from the neutron cross section for temperatures less than the glass temperature (Fig. 7.12). This effect was first demonstrated for Cu–Mn by Ahmed and Hicks (1974). The discrepancy is due to components of the moment which are relaxing more slowly than the time scale of the susceptibility measurement. As can be seen from eqn (2.11) the energy integrated neutron cross section is related to the zero frequency susceptibility plus any truly static components. At higher temperatures the weak variation of the small κ cross section with temperature closely follows the variation of the product of temperature and susceptibility.

Aeppli *et al.* (1985) also tried to characterize the variation of the inelasticity of the scattering with temperature. At $2T_g$ the width of the scattering is about 0.1 meV; again similar to that for Cu–Mn, and varies with temperature roughly as would be expected for an Arrhenius law. A more detailed characterization would need some higher resolution spectroscopy like neutron spin echo.

$Al_2Mn_3Si_3O_{12}$ in an entirely different sort of atomic glass but with spin glass behaviour very similar to the amorphous MnSi. Nägele *et al.* (1979) separated out much of the magnetic scattering by comparison with the scattering from $Al_2Ca_3Si_3O_{12}$ (Fig. 7.13) leaving what can be seen as a quasielastic component in addition to an elastic peak with approximately the instrumental resolution. The latter will contain a nuclear component because of the different scattering lengths of manganese and calcium. The quasielastic scattering can be well fitted by an Arrhenius law above about 15 K (compare. $T_g = 2.4$ K) if the elastic line is

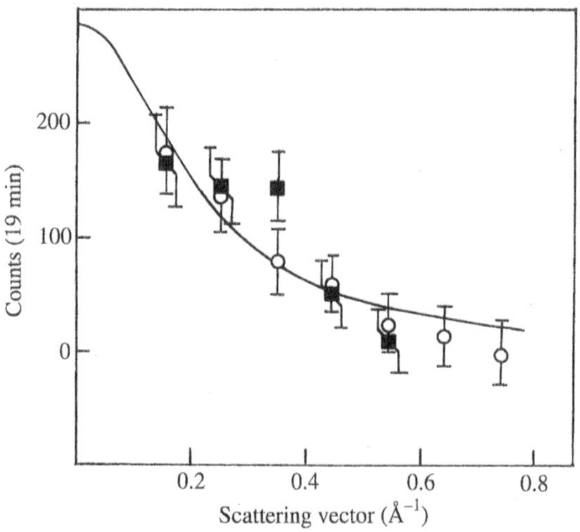

Fig. 7.11 The magnetic scattering from amorphous MnSi at 6 K (○) and 15 K (■). (After Aeppli *et al.* 1985.)

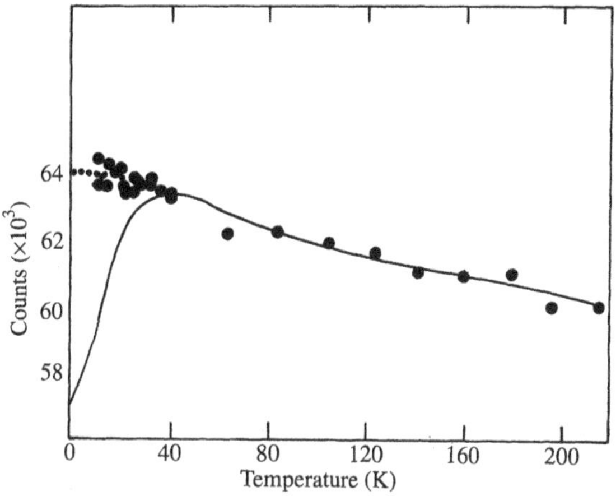

Fig. 7.12 The measured low κ magnetic scattering from amorphous MnSi compared with the d.c. susceptibility multiplied by temperature (solid line), κ = 0.16 Å^{-1}. (After Aeppli *et al.*1985.)

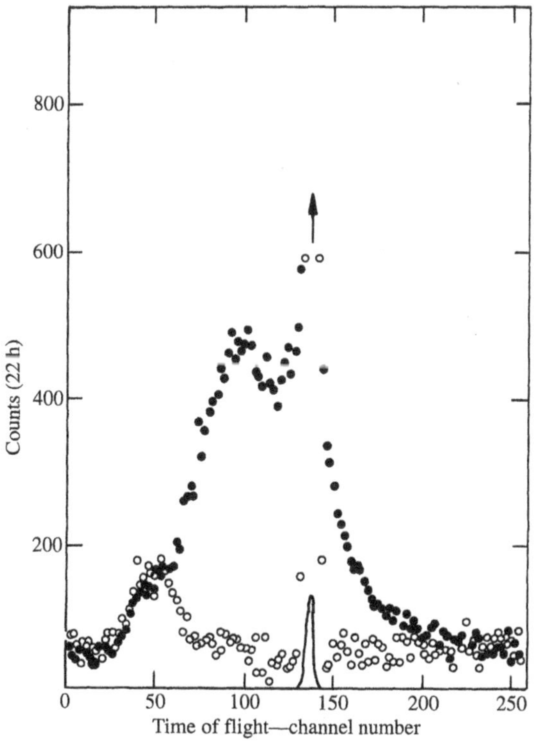

Fig. 7.13 Room temperature time of flight spectra for $Al_2Mn_3Si_3O_{12}$ (●) and $Al_2Ca_3Si_3O_{12}$ (○) glass. The instrumental resolution is shown. 2θ = 61°, E_0 = 3.55 meV. (After Nägele *et al.* 1979.)

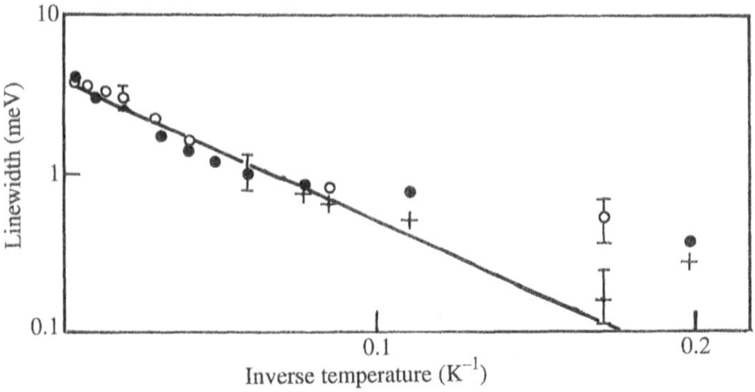

Fig. 7.14 Line width of the quasielastic scattering from $Al_2Mn_3Si_3O_{12}$ with reciprocal temperature for average values of the scattering at $\kappa > 1.2$ Å^{-1}. (After Nägele *et al.* 1979.)

also fitted to a Gaussian. This is shown in Fig. 7.14 where other points below 15 K, also shown, deviate from the Arrhenius law due to fixing the Gaussian width of the elastic line at its room temperature value. From experience with other spin glasses which have been examined with higher resolution spectroscopy there is a good chance that none of the magnetic component of the 'elastic' component is truly elastic.

7.4 AMORPHOUS RE-ENTRANT SPIN GLASSES

This section will concentrate on the metallic glass $(Fe_{1-C}Mn_C)_{75}P_{16}B_6Al_3$ which has a Curie temperature up to about $C = 0.35$ and a spin glass freezing temperature for larger C. From about $C = 0.2$ there is a re-entrant spin glass phase below the ferromagnetic phase in temperature. Figure 7.15 shows the variation of low field magnetization with temperature. The sharp rise of magnetization at the Curie temperature is seen and the sharp decrease of magnetization as the temperature is further decreased can also be seen for compositions with the re-entrant spin glass phase. This behaviour is very similar to several crystalline ferromagnetic re-entrant spin glass alloys systems like Fe_CAu_{1-C} (Murani 1981) and Fe_CCr_{1-C} (Schmatz 1973).

$(Fe_{1-C}Mn_C)_{75}P_{16}B_6Al_3$ shows anomalies in small angle neutron scattering at the same temperatures. The upper one corresponds to the critical scattering at the Curie temperature and the lower one is due to an increasing small angle quasielastic scattering. This small angle scattering was examined by Aeppli *et al.* (1983) who found that it could be fitted by a Lorentzian in κ yielding a correlation length which increased as the temperature was lowered into the ferromagnetic region and then decreased again at still lower temperatures (Fig. 7.16). For compositions away from the multicritical point it was found that at low temperatures a power law in κ or an additional squared Lorentzian term was needed to

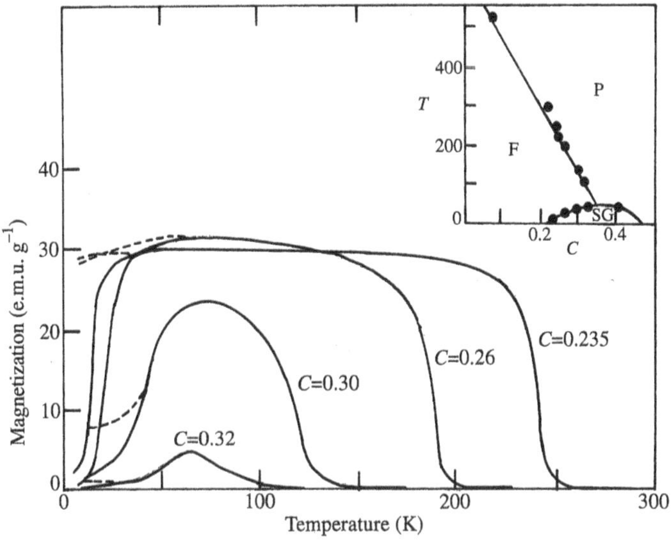

Fig. 7.15 Low field magnetization as a function of temperature for several amorphous Fe–Mn alloys ($(Fe_{1-C}Mn_C)_{75}P_{16}Al_3$). The inset shows the phase diagram. (After Mirebeau *et al.* 1990.)

describe the κ variation. Aeppli *et al.* (1983) attributed the appearance of the Lorentzian squared term as due to random fields from the increasingly static magnetically glassy regions at low temperatures affecting the stability of the ferromagnetic phase. An infinitesimally small random field will destroy long range ferromagnetic order for a short range Heisenberg coupled system in three dimensions. This is an important point which implies that the spontaneous ferromagnetic moment is lost at low temperatures in the re-entrant spin glass phase. It is not simple to determine whether the decrease in magnetization evident in low field magnetization measurements is in fact due to the loss of spontaneous moment. Certainly Hennion *et al.* (1984) would argue from their measurement of magnons in the re-entrant glassy phase for crystalline Ni–Mn that the spontaneous moment is preserved at least over lengths of 150 Å.

One way of determining the length scale over which regions of spontaneous moment exist is by using neutron depolarization through thin films. Figure 7.17 (Mirebeau *et al.* 1990) shows the variation of polarization with temperature through a 60 μm sample. Depolarization starts at the Curie temperature. At lower temperatures the amount of depolarization depends on the magnetic history of the sample, but the irreversibility starts at temperatures much higher than the re-entrant glass temperature indicated by the low field magnetization measurements. Depolarization persists to the lowest temperatures measured, unfortunately for this sample just into the re-entrant glass phase. For $C = 0.30$ the depolarization was measured right into the re-entrant glass phase where a

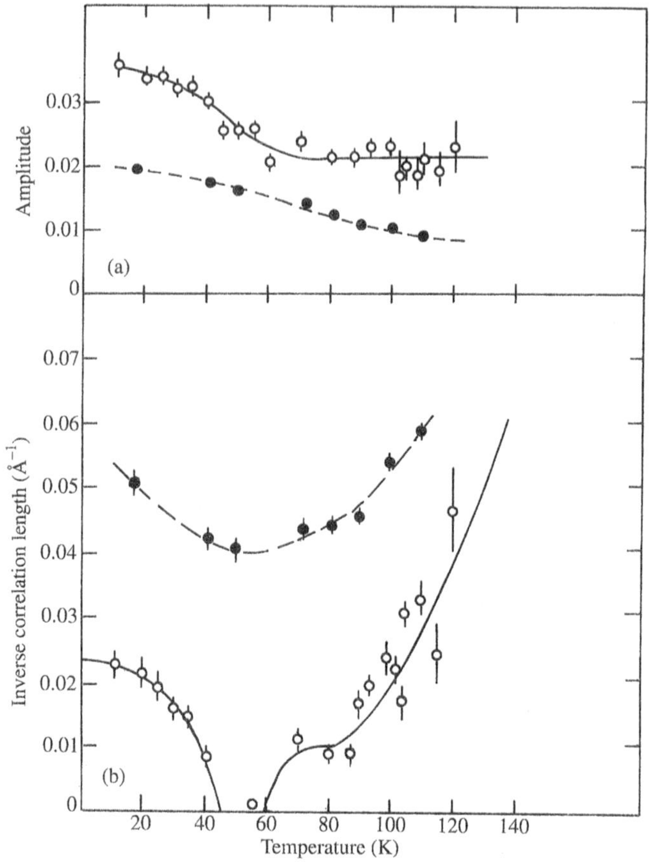

Fig. 7.16 Temperature dependence of (a) the Lorentzian amplitude and (b) the inverse correlation length κ_0 of small angle scattering from amorphous Fe–Mn. (●) $C = 0.35$, (○) $C = 0.32$. (After Aeppli *et al.* 1983.)

domain size of 2000 Å was estimated from the wavelength dependence of the depolarization.

One other aspect of the magnetic structure of this interesting material needs to be discussed, and that is the possible existence of noncollinear components in the moment distribution. Hennion *et al.* (1986) have inferred the presence of transverse moment components by the behaviour of the neutron small angle scattering in a field. Figure 7.18 shows the change in the small angle scattering with applied field. At low fields the scattering decreases monotonically with scattering vector but at higher fields a peak appears at larger and larger scattering vectors indicating that considerable redistribution of the magnetic moment is taking place even though the sample is reasonably well saturated. This behaviour is similar to that observed for $TbFe_2$ ferromagnetic glass (Spano and Rhyne 1986)

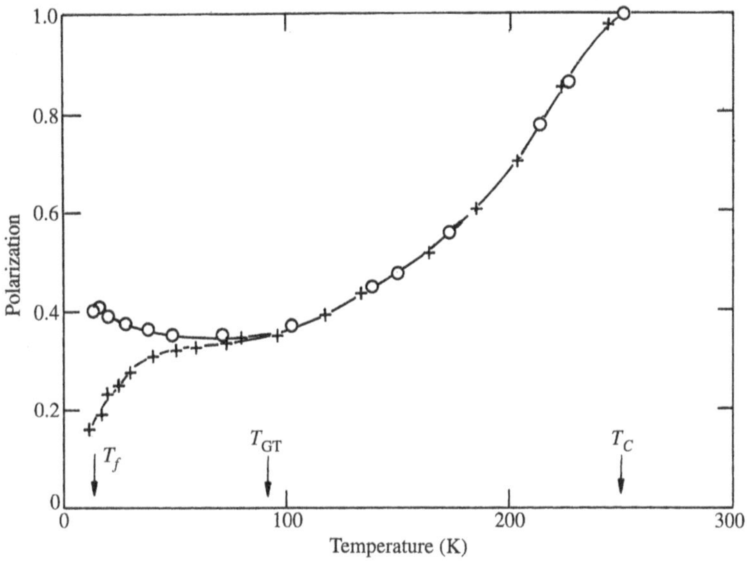

Fig. 7.17 The depolarization in amorphous $(Fe_{1-c}Mn_c)_{75}P_{16}B_6Al_3$ for $C = 0.235$; $H = 560$ A m^{-1}, $\lambda = 5$ Å. (After Mirebeau *et al.* 1990.)

including the observation that the scattering is more intense when the scattering vector is in the direction of the magnetic field. It is this last observation which Hennion *et al.* (1986) use to conclude that this intensity is associated with moment components transverse to the ferromagnetic direction. This certainly could be inferred if the whole moment distribution were rigidly attached to the ferromagnetic component, but Rhyne *et al.* attribute their similar results to a component of the moment which is free to rotate in laboratory times with a consequent relative enhancement of the scattering in the field direction relative to that perpendicular to the field (Lovesey, 1984, p.9). In the light of the the polarization analysis results on $Fe_{83}B_{17}$ (Fischer 1988) and $(Fe_CNi_{1-C})_{78}B_{12}Si_{10}$ (Cowley *et al.* 1992) it extremely likely that the Fe–Mn metalloid glass also has transverse moment components, but it will need a polarization analysis experiment to confirm it.

The inelastic spectra of $(Fe_{1-C}Mn_C)_{75}P_{16}B_6Al_3$ consist of a central quasielastic peak with magnon sidebands. The magnons get increasingly hard to measure as the temperature is decreased and the central quasielastic peak increases in intensity and very high energy resolution is required. Figure 7.19 shows the results of Hennion *et al.* (1986) for $C = 0.23$ and for a series of temperatures. The magnon shoulders are still clearly seen on the very intense central peak at 10 K. Figure 7.20 shows the variation of magnon stiffness with temperature for two different fittings of the lineshapes, firstly according to a damped harmonic oscillator and then according to a damped Lorentzian. Both produce a magnon stiff-

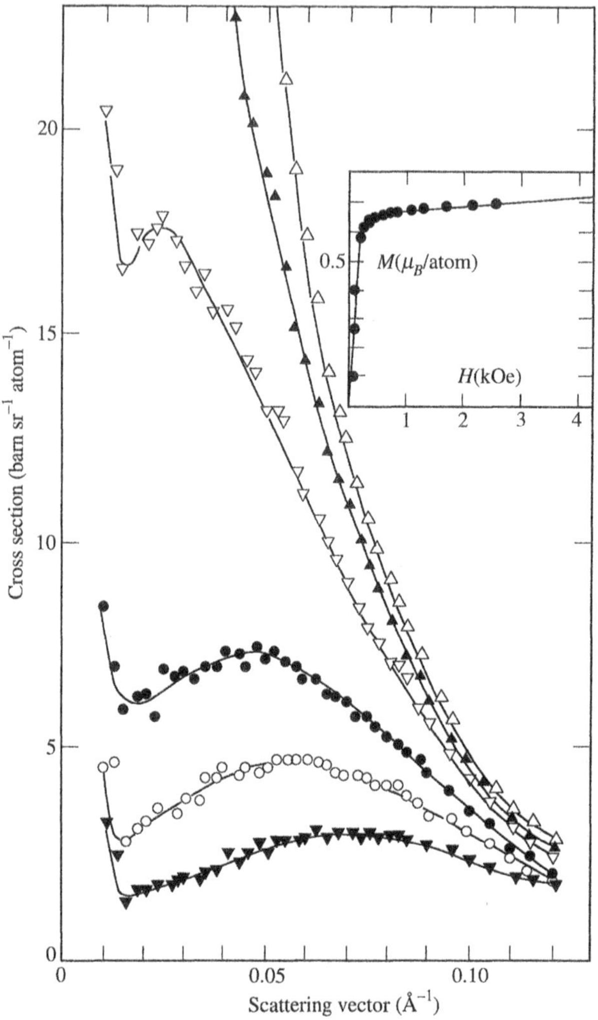

Fig. 7.18 The small angle scattering intensity from amorphous $(Fe_{1-c}Mn_c)_{75}P_{16}B_6Al_3$ for C = 0.235 as a function of field parallel to κ. (Δ) 64 kA m^{-1}, (▲) 144 kA m^{-1}, (∇) 208 kA m^{-1}, (●) 336 kA m^{-1}, (○) 464 kA m^{-1}, (▼) 624 kA m^{-1}. The insert shows magnetization versus field. (After Hennion *et al.* 1986.)

ness which after increasing below the Curie temperature reduces at lower temperatures before rising again at the lowest temperatures. This is in contrast to the conclusions drawn from an earlier lower resolution study (Aeppli *et al.* 1984) in which magnons were not visible at low temperatures. It was then thought that the spontaneous moment had been destroyed presumably by the random fields and that ferromagnetic magnons were not supported.

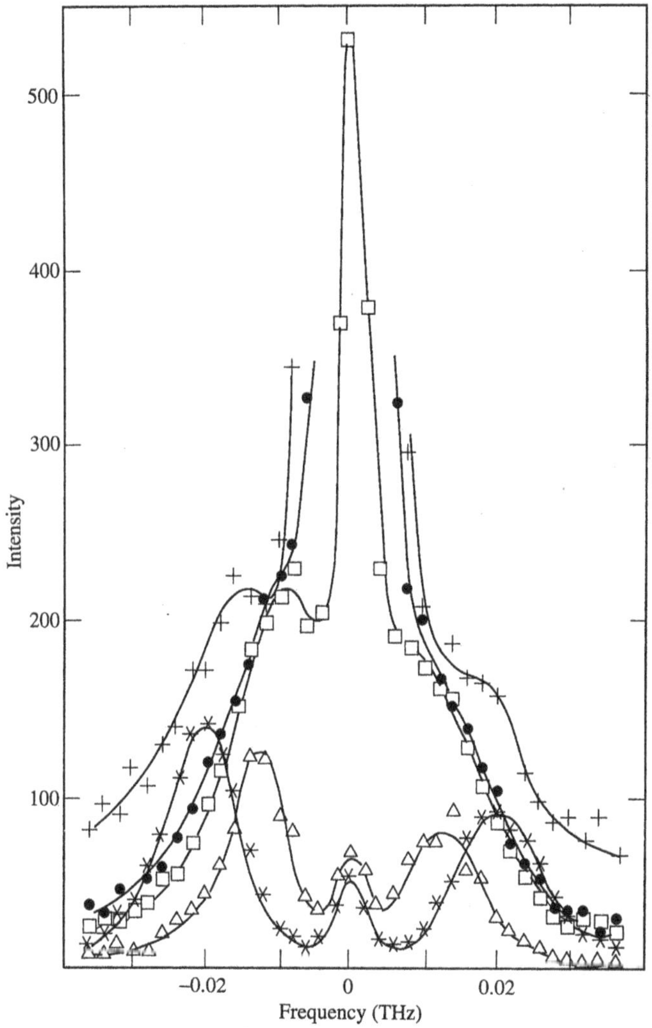

Fig. 7.19 The evolution with temperature of inelastic scattering from $(Fe_{1-c}Mn_c)_{75}P_{16}Al_3$ for $C = 0.235$, zero field and $\kappa = 0.046$ Å^{-1}. M is the monitor count. (+) $T = 10$ K, $M = 4500$; (●) $T = 30$ K, $M = 1500$; (□) $T = 60$ K, $M = 7500$; (*) $T = 140$ K, $M = 3200$; (Δ) $T = 230$ K, $M = 2000$. (After Hennion *et al.* 1986.)

Here, as for simpler systems, neutron scattering is shown to be a powerful probe of magnetic properties. The short range nature of the structure of amorphous materials is definitively characterized by neutron diffraction, and the dynamic behaviour can be made apparent in careful experiments. The limitation of neutron diffraction for these purposes lies in its sensitivity. All neutron sources are low intensity when compared with other particle probes of materials,

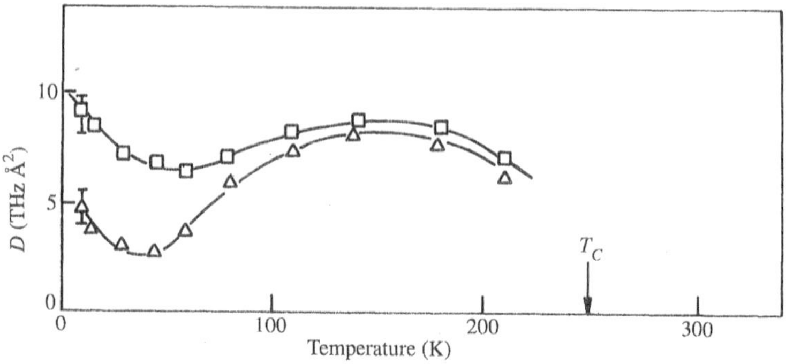

Fig. 7.20 The magnon stiffness constant in $(Fe_{1-c}Mn_c)_{75}P_{16}B_6Al_3$ fitted with a damped harmonic oscillator (□) or a double Lorentzian (Δ) line shape. (After Hennion *et al.* 1986.)

so that definitive experiments can be time consuming. Nevertheless, the results of neutron experiments have always been significant because of their directness and because the spatial and time domain which they probe is crucial for the understanding of complex magnetic materials.

References

Aeppli, G., Shapiro, S.M., Birgeneau, R.J., and Chen, H.S. (1983). *Phys. Rev. B* **28**, 5160.

Aeppli, G., Shapiro, S.M., Birgeneau, R.J., and Chen, H.S. (1984). *Phys. Rev. B* **29**, 2589.

Aeppli, G., Hauser, J.J., Shirane, G., and Uemura, Y.J. (1985). *Phys. Rev. Lett.* **54**, 843.

Ahmed, N. and Hicks, T.J. (1974). *Solid State Comm.* **15**, 415.

Bletry, J. and Sadoc, J.F. (1975). *J. Phys. F* **5**, L110.

Cowley, R.A., Cowlam, N. and Cussen, L.D. (1988). *J. de Phys.* C8, 1285.

Cowley, R.A., Cowlam, N., Ivison, P.K. and Martinez, J. (1992). *J. Mag. Mag. Mat.* **104–107**, 159.

Faber, T.E. and Ziman, J.M. (1965). *Phil. Mag.* **11**, 153.

Fernandez-Baca, J.A., Lynn, J.W., Rhyne, J.J., and Fish, G.E. (1987). *Phys. Rev. B* **36**, 8497.

Fischer, K.H. (1988). *J. de Phys.* **C8**, 1293.

Hennion, B., Hennion, M., Hippert, F., and Murani, A.P. (1984). *J. Phys. F* **14**, 489.

Hennion, B., Hennion, M., and Mirebeau, I. (1986). *Physica B&C* **6**, 49.

Hennion, M., Mirebeau, I., Hennion, B., and Lequien, S. (1986). *Europhys. Lett.* **2**, 393.

Ishikawa, Y., Yamada, K., Tajima, K., and Fuchamichi, K. (1981). *J. Phys. Soc. Jap.* **50**, 1958.

Lovesey, S.W. (1984). *Theory of neutron scattering from condensed matter*. Vol. 2. Oxford University Press.

Mirebeau, I., Itoh, S., Mitsuda, S., Watanabe, N., Endoh, Y., Hennion, M. *et al.* (1990). *Phys. Rev. B* **41**, 11405.

Mook, H.A. and Lynn, J.W. (1984). *Phys. Rev. B* **29**, 4056.

Mook, H.A. and Tsuei, C.C. (1977). *Phys. Rev. B* **16**, 2184.

Moon, R.M. (1971). *Int.J.Mag.* **1**, 219.

Murani, A.P. (1981). *J. Mag. Mag. Mat.* **22**, 271.

Nägele, W., Blanckenhagen, P.von, Knorr, K., and Suck, J.B. (1979). *Z. Physik B* **33**, 251.

Paul, D.McK., Cowley, R.A., Stirling, W.G., Cowlam, N., and Davies, H.A. (1982). *J. Phys. F* **12**, 2687.

Schmatz, W. (1973). X-ray and neutron scattering studies on disordered crystals. In *Treatise on materials science and technology,* (ed. H. Herman) p. 105. Academic Press, New York.

Shirane, G., Axe, J.D., Majkrzak, C.F., and Mizoguchi, T. (1982). *Phys. Rev. B* **26**, 2575.

Spano, M.L. and Rhyne, J.J. (1986). *J. Mag. Mag. Mat.* **54–57**, 93.

Stusser, N., Rekveldt, M.Th., and Spruijt, T. (1979). *J. Appl. Phys.* **50**, 2122.

Tarvin, J.A., Shirane, G., Birgeneau, R.J. and Chen, H.S. (1978). *Phys. Rev.* B **17,** 241.

Wu, Guoan., Cowlam, N., Davies, H.A., Cowley, R.A., Paul, D.McK., and Stirling, W.G. (1982). *J. de Phys.* **C7**, 71.

Ze Xianyu, Ishikawa, Y., Fukunaga, T., and Watanabe, N. (1985). *J. Phys. F* **15**, 1799.

Name index

Abrahams, S.C. 72
Aeppli, G. 132–4, 136, 138
Aharony, A. 132
Ahmed, N. 88–91, 99
Aldred, A.T. 19, 33–4, 42, 44
Alloul, H. 98
Anderson, P.W. 98
Arrott, A. 89

Bacon, G.E. 65, 89
Bakshi, E. 67–70
Balcar, E. 27, 36
Belanger, D.P. 82–4
Bhatt, R.N. 107
Birgeneau, R.J. 81–3
Bletry, J. 122–4
Blume, M. 3, 9–10, 101
Buyers, W.J.L. 75–80

Cable, J.W. 19–22, 34–7, 42–3, 45, 92–4
Campbell, I.A. 23
Child, H.R. 42–3, 45
Clapp, P. 46
Coles, B.R. 103–4
Collins, M.F. 17, 20, 22–4
Comly, J.B. 19, 21
Cowley, R.A. 75–7, 79, 122, 125–6, 137
Cywinski, R. 65

Dagleish, P.A. 109
Davis, J.R. 54, 56–7, 92

Edwards, S.F. 98

Faber, T.E. 122
Felsch, W. 99–100, 104–5
Fernandez-Baca, J.A. 130
Fischer, K.H. 137

Garland, J.W. 31
Gibbs, P. 107
Giebultowicz, T. 100, 113–14
Gomès, A.A. 23
Gray, E.M. 93
Grinstein, G. 107

Harders, T.M. 93–4
Hatherley, M. 72
Hennion, B. 115–17, 135–6, 138
Hennion, M. 73–4, 137, 139–40
Hicks, T.J. 21–2, 24, 34, 36–7, 39–45, 56–65, 67–70, 88–91, 99
Holden, T.M. 22, 25, 100
Hutchings, M.T. 81

Ishikawa, Y. 130

Kajzar, F. 23
Kasuya, T. 99
Kawarazaki, S. 63
Kennedy, S.J. 63, 65
Kouvel, J.S. 89

Leoni, F. 47
Lequien, S. 117
Lindley, L. 100
Lovesey, S.W. 2–3, 10, 66–7, 69, 137
Low, G.G. 17, 20–5
Lowde, R.D. 72
Lynn, J.A. 127–9

Maier, B. 51
Maletta, H. 99–100, 104–5
Marshall, W. 27, 29, 32–3, 36, 54, 55, 60–7, 69
Medina, R.A. 20, 31, 34–7
Meneghetti, D. 65, 89
Menzinger, F. 48
Mezei, F. 18, 98–9, 112–13
Mirebeau, I. 106, 135, 137
Mook, H.A. 127–9
Moon, R.M. 123
Morgownik, A.F.J. 91
Moss, S.C. 46

Moze, O. 41, 45, 58–64
Mukamel, D. 107
Murani, A. 92, 95, 97–9, 103, 114–15, 134
Mydosh, J.A. 91

Nägele, W. 132–4
Nakai, Y. 65

Owen, J. 89

Parette, G. 23
Paul, D. McK. 131

Rainford, B.D. 64, 66
Rhyne, J.J. 136

Sacchetti, F. 47
Sadoc, J.F. 122–4
Sato, H. 89
Schärpf, O. 52
Scheuer, H. 113
Schmatz, W. 134
Shapiro, S.M. 104–5, 118
Shirane, G. 131
Shull, C.G. 31, 67
Sidhu, S.S. 65, 89
Smith, J.H. 89, 107
Soukoulis, C.M. 96
Soven, P. 46
Spano, M.L. 136
Staunton, J.B. 27
Steigenberger, U. 100–2
Stusser, N. 128

Tarvin, J.A. 129–30
Tholence, J.L. 97–8
Tsuei, C.C. 131
Tsunoda, Y. 65, 92

Uchishiba, H. 58

Wells, P. 89, 93
Werner, S.A. 89, 92
Wilkinson, M.K. 31
Wong, Po-zen 101–3, 107–8, 113
Wu, Guoan 124–5

Young, A.P. 107

Ze Xianyu 124
Ziman, J.M. 122

Subject index

$Al_2Mn_3Si_3O_{12}$ glass 132–4
Arrhenius 112, 117, 134
Au–Fe alloys 92, 103–4
  magnons in re-entrant spin glass 114–15
average
  configurational 18, 46, 77, 87–8
  over sites 6
  thermal 4, 6, 15, 28, 87

Born approximation 2
Bragg cutoff 17

$Cd_{1-C}Mn_CTe$ 100
  magnons 113–14
clustering, atomic 65
cobalt alloys, dilute
  Co–Cr 22
  Co–Mn 22
  Co–Mo 21
  Co–Ni 22
  Co–V 22
coherent potential approximation 27, 46–8, 77–9
Cowley short range order parameters 28–31, 58–60, 87–9
$Co_{0.7}Mn_{0.3}F_2$, magnons 75–6, 79
$Co_{83}P_{17}$ glass 122–4
$Co_{1-C}Zn_CF_2$ 79
Cr–Fe
  magnons in re-entrant spin glass 117
critical concentration for ferromagnetism 35–6, 104
critical exponents 80–3
critical scattering 104–5
cross section
  dependence on initial polarization 5
  isolation of magnetic 5, 16
  at $\kappa = 0$ 19, 29–31, 32, 39, 89–90, 98
  unanalysed 4
Cu–Mn alloys
  spin echo results 111–12

D7 diffractometer 51–2
density of electron states 46
density matrix 3

dispersion relation, magnon 71–2, 75, 77
domains, ferromagnetic 106–7

$Eu_{0.4}Sr_{0.6}S$ 99–100
  spin echo data 113
$Eu_{0.54}Sr_{0.46}S$ 118
$Eu_CSr_{1-C}S$ 99, 104–6, 117–18
exchange field 37, 56
exchange interaction 71–2, 88, 91–2
  anisotropic 100

$Fe_{83}B_{17}$ glass 124
$Fe_{86}B_{14}$ glass 127
FeF_2, critical scattering 81
$Fe_{0.4}Mg_{0.6}Cl_2$
  magnetic excitations 113–14
$Fe_{0.55}Mg_{0.45}Cl_2$ 107–8
$Fe_{1-C}Mg_CCl_2$ 101
$(Fe_{1-C}Mn_C)_{75}P_{16}B_6Al_3$ glass
  inelastic scattering 137, 139–40
  neutron depolarization 135, 137
  re-entrant spin glass properties 134–40
  small angle scattering 136, 138
Fe–Ni, concentrated alloys
  magnons 72–3
$Fe_{40}Ni_{40}P_{14}B_6$ 127–8
$(Fe_{65}Ni_{35})_{75}P_{16}B_6Al_3$ 127–8
$Fe_{75}P_{15}C_{10}$ glass 131
Fermi chopper 52
Fermi surface 93
Fe–V 27
$Fe_{0.46}Zn_{0.54}F_2$, critical scattering 83–4
$Fe_{0.5}Zn_{0.5}F_2$, critical scattering 81–3
flipper chopper 52

Green function 77–8
  relationship with susceptibility 77

Harwell 17
Heisenberg system 71, 107

interference between neutron spin states 11
iron alloys, dilute

iron alloys, dilute (*cont.*)
  Fe–Cr 23
  Fe–Cu 23
  Fe–Ir 23
  Fe–Mn 23
  Fe–Mo 23
  Fe–Os 23
  Fe–Pd 23–4
  Fe–Pt 23
  Fe–Re 23
  Fe–Rh 23
  Fe–Ru 23
  Fe–Ti 23
  Fe–V 23
  Fe–W 23
iron, face centred cubic 92
Ising model 80–1, 83, 107

$KCo_{0.2}Mn_{0.8}F_2$ 77
Kramers–Krönig 9–10, 111

lattice distortion 68, 96–7
LONGPOL 52–4

magnetic dipolar interactions 107
magnetic form factor 15, 88
magnetic induction 2
magnetic moment
  operator for total 3
magnetization
  distribution 3
magnons
  from spin echo data 118
  in $Cd_{1-C}Mn_CTe$ 100
magnon stiffness 72–3
  in ferromagnetic glasses 127–9
Marshall model 31–6, 54–5, 60–1, 63, 68–9
Metglas 127–8
Mn–Cu alloy system 107
Mn–Cu, concentrated alloys 65–6
Mn–Fe, concentrated alloys 65
Mn–Ni, concentrated alloys 58–65
MnSi glass
  spin glass properties 131–3
$Mn_{0.68}Zn_{0.32}F_2$ 67–9
$Mn_{1-C}Zn_CF_2$ 67, 70, 79–80
molecular field, local 88

neutron
  depolarization 106–7, 135, 137
  gyromagnetic ratio 2
  magnetic moment operator 2
  nuclear interaction 2
  spatial states 2–3
  spin states 3–4
neutron spin echo 98
nickel alloys, dilute
  Ni–Al 20
  Ni–Cr 20–1
  Ni–Cu 19
  Ni–Fe 20–1
  Ni–Ga 20
  Ni–Ge 20
  Ni–Mn 20–1
  Ni–Mo 20
  Ni–Nb 20
  Ni–Re 20
  Ni–Rh 20–1
  Ni–Ru 20
  Ni–Sb 20
  Ni–Si 20
  Ni–Sn 20
  Ni–V 20–1
  Ni–W 20
  Ni–Zn 19–20
Ni–Cr, concentrated alloy 39
Ni–Cu, concentrated alloy 33–6, 39–41, 47–8
Ni–Mn
  concentrated alloy 41, 45
  magnons 72–3
  magnons in re-entrant spin glass 115–17
  re-entrant spin glass 135
Ni–Pd, concentrated alloy 41–5
Ni–Rh, concentrated alloy 41
non collinear antiferromagnet 63–5
nuclear spin 5–6
  operator 6
nuclear-magnetic cross term 16
  isolation 16

orbital magnetic moment 2

palladium alloys, dilute
  Pd–Co 25
  Pd–Fe 24–5
Pauli spin matrix 2, 11
percolation concentration 104
perturbation theory 7
PLUTO reactor 17
polarization 121–2
  dependence on initial 10
  final 4, 8–9, 12, 14
polarization analysis 50–4, 58, 65, 67, 86, 89, 92, 94, 98, 121

quasielastic scattering 80, 110
  magnetic 9–10

random fields 83, 138
relaxation
  ln *t* 112
  non-exponential 98, 112

relaxation time 98, 111, 118
resolution
  angular 17
  wavelength 17

small angle scattering 93, 95, 103–6, 136, 138
spin density wave 89
spin flipper 53
  psuedo random sequence 53–4, 67
spin–spin correlation 87, 98
  temperature variation of length 101
spiral structure, magnetic 5, 14
supermirrors 52
susceptibility
  a.c. 98–9, 101–2
  comparison with cross section 10
  dynamic in ferromagnetic glasses 130–1
  dynamic frequency dependent 7–9
  generalized 110
  initial atomic 37, 56
  real and imaginary parts 6–9
  from spin echo measurements 111–12
  zero frequency dynamic 80–2, 88

thermal fluctuations 16
time of flight spectroscopy 53–4, 109
transition amplitude 3
  spin 12
transition probability
  between spin states 3, 12–13
transverse moment
  in antiferromagnetic Mn alloys 62–6
  in ferromagnetic glasses 122, 125–6
triple axis spectrometer 50, 109